FM 23-35

US ARMY FIELD MANUAL

PISTOLS

AND

REVOLVERS

1946

WORLD WAR II
CIVILIAN REFERENCE EDITION
UNABRIDGED TECHNICAL MANUAL ON VINTAGE AND COLLECTIBLE
SIDE AND HANDHELD FIREARMS FROM THE WARTIME ERA

U.S. WAR DEPARTMENT

Doublebit Press

New content, introduction, cover design, and annotations Copyright © 2020 by Doublebit Press. All rights reserved.

Doublebit Press is an imprint of Eagle Nest Press
www.doublebitpress.com
Cherry, IL, USA

Original content under the public domain; unrestricted for civilian distribution. Originally published in 1946 by the U.S. War Department.

This title, along with other Doublebit Press books are available at a volume discount for youth groups, clubs, or reading groups. Contact Doublebit Press at info@doublebitpress.com for more information.

Military Outdoors Skills Series: Volume 7

Doublebit Press Civilian Reference Edition ISBNs
Hardcover: 978-1-64389-156-9
Paperback: 978-1-64389-157-6

Doublebit Press, or its employees, authors, and other affiliates, assume no liability for any actions performed by readers or any damages that might be related to information contained in this book. Some of the material in this book may be outdated by modern standards. This text has been published for historical study and for personal literary enrichment. Remember to be safe with any activity that you do in the outdoors and to help do your part to preserve and be a good steward of our great American wild lands.

The Military Outdoors Skills Series
Historic Field Manuals and Military Guides
on Outdoors Skills and Travel

Military manuals contain essential knowledge about outdoors life, thriving while in the field, and self-sufficiency. Unfortunately, many great military books, field manuals, and technical guides over the years have become less available and harder to find. These have either been rescinded by the armed forces or are otherwise out of print due to their age. This does not mean that these manuals are worthless or "out of date" – in fact, the opposite is true! It is true that the US Military frequently updates its manuals as its protocols frequently change based on the current times and combat situations that our armed services face. However, the knowledge about the outdoors over the entire history of military publications is timeless!

By publishing the **Military Outdoors Skills Series**, it is our goal at Doublebit Press to do what we can to preserve and share valuable military works that hold timeless knowledge about outdoors life, navigation, and survival. These books include official unrestricted texts such as army field manuals (the FM series), technical manuals (the TM series), and other military books from the Air Force, Navy, and texts from before 1900. Through remastered reprint editions of military handbooks and field manuals, outdoors enthusiasts, bushcrafters, hunters, scouts, campers, survivalists, nature lore experts, and military historians can preserve the time-tested skills and institutional knowledge that was learned through hard lessons and training by the U.S. Military and our expert soldiers.

Soldiers were the original campers and survivalists! Because of this, military field manuals about outdoors life contain essential knowledge about thriving in the wilds. This book is not just for soldiers!

This book is an important contribution to outdoors literature and has important historical and collector value toward preserving the American outdoors tradition. The knowledge it holds is an invaluable reference for practicing skills related to thriving in the outdoors. Its chapters thoroughly discuss some of the essential building blocks of outdoors knowledge that are fundamental but may have been forgotten as equipment gets fancier and technology gets smarter. In short, this book was chosen for Historic Edition printing because much of the basic skills and knowledge it contains could be forgotten or put to the wayside in trade for more modern conveniences and methods.

Although the editors at Doublebit Press are thrilled to have comfortable experiences in the woods and love our high-tech and light-weight equipment, we are also realizing that the basic skills taught by the old experts are more essential than ever as our culture becomes more and more hooked on digital technology. We don't want to risk forgetting the important steps, skills, or building blocks involved with thriving in the outdoors. This Civilian Reference Edition reprint represents a collection of military handbooks and field manuals that are essential contributions to the American outdoors tradition despite originating with the military. In the most basic sense, these books are the collection of experiences by the great experts of outdoors life: our countless expert soldiers who learned to thrive in the backwoods, deserts, extreme cold environments, and jungles of the world.

With technology playing a major role in everyday life, sometimes we need to take a step back in time to find those basic building blocks used for gaining mastery – the things that we have luckily not completely lost and has been recorded in books over the last two centuries. These skills aren't forgotten, they've just been shelved. *It's time to unshelve them once again and reclaim the lost knowledge of self-sufficiency.*

Based on this commitment to preserving our outdoors heritage, we have taken great pride in publishing this book as a complete original work. We hope it is worthy of both study and collection by outdoors folk in the modern era of outdoors and traditional skills life.

Unlike many other photocopy reproductions of classic books that are common on the market, this Historic Edition does not simply place poor photography of old texts on our pages and use error-prone optical scanning or computer-generated text. We want our work to speak for itself, and reflect the quality demanded by our customers who spend their hard-earned money. With this in mind, each Historic Edition book that has been chosen for publication is carefully remastered from original print books, *with the Doublebit Civilian Reference Edition printed and laid out in the exact way that it was presented at its original publication.* We provide a beautiful, memorable experience that is as true to the original text as best as possible, but with the aid of modern technology to make as beautiful a reading experience as possible for books that are typically over a century old. Military historians and outdoors enthusiasts alike are sure to appreciate the care to preserve this work!

Because of its age and because it is presented in its original form, the book may contain misspellings, inking errors, and other print blemishes that were common for the age. However, these are exactly the things that we feel give the book its character, which we preserved in this Historic Edition. During digitization, we ensured that each illustration in the text was clean and sharp with the least amount of loss from being copied and digitized as possible. Full-page plate illustrations are presented as they were found, often including the extra blank page that was often behind a plate. For the covers, we use the original cover design to give the book its original feel. We are sure you'll appreciate the fine touches and attention to detail that your Historic Edition has to offer.

For outdoors and military history enthusiasts who demand the best from their equipment, the Doublebit Press Civilian Reference Edition reprint of this military manual was made with you in mind. Both important and minor details have equally both been accounted for by our publishing staff, down to the cover, font, layout, and images. It is the goal of Doublebit Civilian Reference Edition series to preserve outdoors heritage, but also be cherished as collectible pieces, worthy of collection in any outdoorsperson's library and that can be passed to future generations.

WAR DEPARTMENT FIELD MANUAL
FM 23-35

This manual supersedes FM 23-35, 30 April 1940, including **C 1,** 23 January 1942, and C 2, 6 November 1942; and FM 23-36, 20 October 1941, including C 1, 6 November 1942, and C 2, 17 April 1943

PISTOLS AND REVOLVERS

WAR DEPARTMENT   *JUNE 1946*

United States Government Printing Office

Washington   1946

WAR DEPARTMENT
Washington 25, D. C., 10 June 1946

FM 23-35, Pistols and Revolvers, is published for the information and guidance of all concerned:

[AG 300.7 (4 Apr 46)]

By order of the Secretary of War:

DWIGHT D. EISENHOWER
Chief of Staff

Official:
EDWARD F. WITSELL
*Major General
The Adjutant General*

Distribution:
AAF (10); AGF (40); ASF (2); AAF Comds (5); T (5); Arm & Sv Bd (1); SvC (3); FC (2); Sp Sv Sch (25); USMA (25); Tng C (5); A (5); CHQ (5); D (5); R 2, 4–7, 17, 18, 44 (3); Bn 2–7, 9–11, 17–19, 44, 55 (3); C 2–7, 9–11, 17–19, 44, 55 (2); AF (5); W (5); G (3); S (2); Special distribution.

Refer to FM 21–6 for explanation of distribution formula.

CONTENTS

	Paragraphs	Page
PART ONE. AUTOMATIC PISTOL, CALIBER .45.		
CHAPTER 1. MECHANICAL TRAINING.		
Section I. Description	1–2	1
II. Disassembling and assembling	3–4	6
III. Care and cleaning	5–11	15
IV. Functioning	12–14	20
V. Spare parts and accessories	15–16	25
VI. Ammunition	17–23	26
VII. Individual safety precautions	24–25	30
CHAPTER 2. MANUAL OF THE PISTOL, DISMOUNTED AND MOUNTED.		
Section I. General	26	33
II. Dismounted	27–35	34
III. Mounted	36–42	39
PART TWO. REVOLVER, CALIBER .45.		
CHAPTER 1. MECHANICAL TRAINING.		
Section I. Description	43–44	41
II. Disassembling and assembling	45–48	46
III. Care and cleaning	49–55	60
IV. Functioning	56–59	65
V. Spare parts and accessories	60–61	74
VI. Ammunition	62	75
VII. Individual safety precautions	63–64	75
CHAPTER 2. MANUAL OF THE REVOLVER, DISMOUNTED AND MOUNTED.		
Section I. General	65	77
II. Dismounted	66–70	78
III. Mounted	71–72	81

Paragraphs. Page

PART THREE. MARKSMANSHIP.

CHAPTER 1. KNOWN-DISTANCE TARGETS.

	Paragraphs	Page
Section I. General	73–74	82
II. Preparatory training	75–86	86
III. Courses to be fired	87–90	132
IV. Range practice	91–96	136
V. Known-distance targets and ranges	97–99	156
VI. Small-bore practice	100–104	162

CHAPTER 2. COMBAT FIRING.

	Paragraphs	Page
Section I. General	105	164
II. Dismounted	106–112	165
III. Mounted	113–121	174
IV. Combat firing targets	122–123	187
V. Small-bore practice	124–126	188

CHAPTER 3. ADVICE TO INSTRUCTORS.

	Paragraphs	Page
Section I. General	127–128	189
II. Mechanical training	129–134	191
III. Manual of the pistol and revolver	135	194
IV. Marksmanship	136–139	195

INDEX 199

This manual supersedes FM 23–35, 30 April 1940, including C 1, 23 January 1942, and C 2, 6 November 1942; and FM 23–36, 20 October 1941, including C 1, 6 November 1942, and C 2, 17 April 1943.

PART ONE
AUTOMATIC PISTOL, CALIBER .45

CHAPTER 1

MECHANICAL TRAINING

Section I

DESCRIPTION

1. GENERAL. a. The automatic pistols, caliber .45, M1911 and M1911A1, are recoil-operated, magazine-fed, self-loading hand weapons (figs. 1 and 2). The gas generated in a cartridge fired in the pistol is utilized to perform the functions of extracting and ejecting the empty cartridge case, cocking the hammer and forcing the slide to the rearmost position, thereby compressing the recoil spring. The action of the recoil spring forces the slide forward, feeding a live cartridge from the magazine into the chamber, leaving the weapon ready to fire again.

b. The M1911A1 pistol is a modification of the M1911 pistol. The operation of both models of pistols is exactly the same. The changes consist of the following (fig. 2):

(1) The tang of the grip safety is extended better to protect the hand.

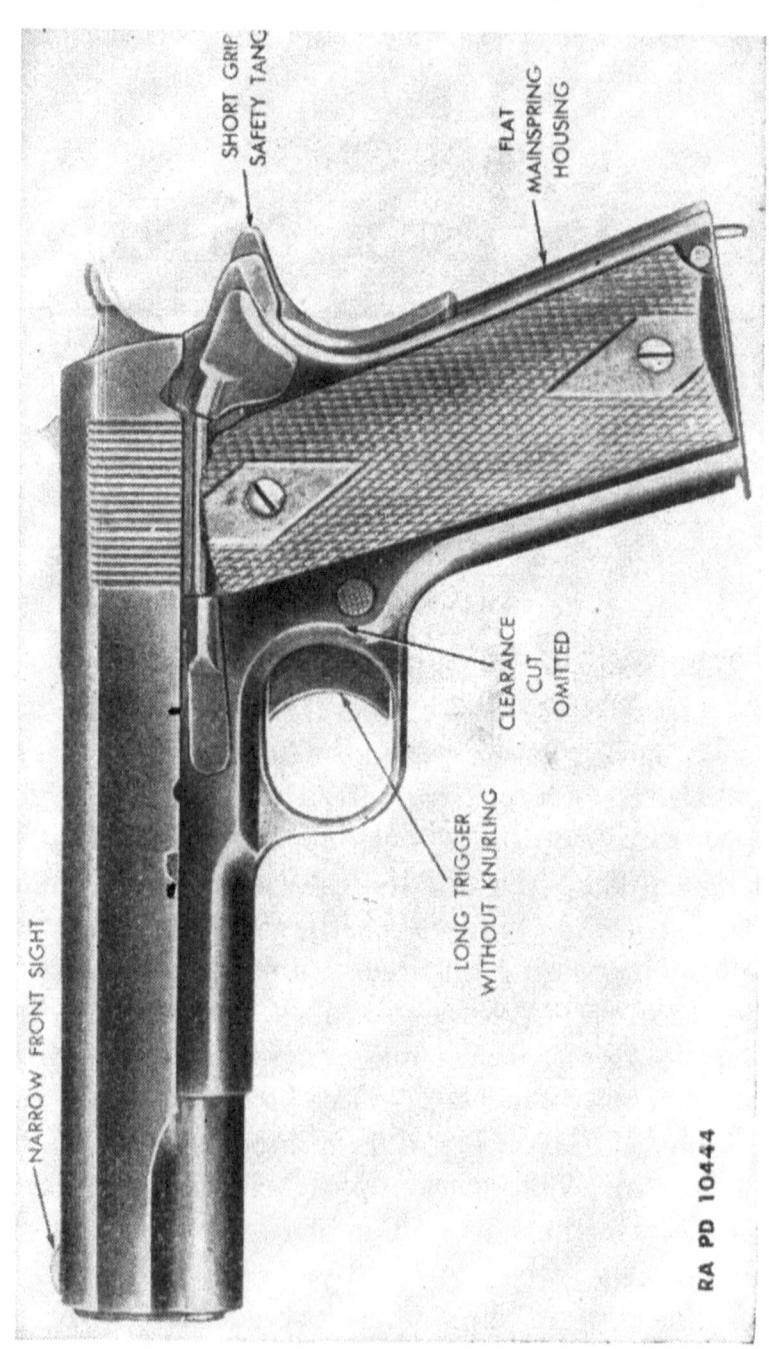

Figure 1. Left side of pistol, M1911.

(2) A clearance cut is made on the receiver for the trigger finger.

Note. For military terms not defined in this manual, see TM 20-205; for list of training publications, see FM 21-6; for training films, film strips, and film bulletins, see FM 21-7; and for training aids, see FM 21-8.

(3) The face of the trigger is cut back and knurled.

(4) The mainspring housing is raised in the form of a curve to fit the palm of the hand and is knurled.

(5) The top of the front sight is widened.

c. The pistol is designed to fire cartridge, ball, caliber .45, M1911. The magazine holds seven cartridges. The upper cartridge is stripped from the magazine and forced into the chamber by the forward motion of the slide. The pistol fires but once at each squeeze of the trigger. When the last cartridge in the magazine has been fired the slide remains open. The magazine catch is then depressed and the empty magazine falls out. A loaded magazine is then inserted, making seven more shots available.

d. The rate of fire is limited by the dexterity of the operator in inserting magazines into the pistol and the ability of the firer to aim and squeeze.

2. GENERAL DATA. a. Dimensions.

(1) *Barrel.*

Caliber of bore_____inches__	0.45
Number of grooves_____	6
Twist in rifling, uniform L. H., one turn in _____inches__	16
Length of barrel_____do____	5.03

(2) *Pistol.*

    Overall length of pistol_____inches__ 8.593
    Height of front sight above axis of
      bore _____inches__ 0.5597

b. Weights.

    Weight of pistol with magazine
                                  pounds__ 2.437
    Weight of loaded magazine, 7 rounds
      (approximate)_____pounds__ 0.481
    Weight of empty magazine____do____ 0.156

c. Trigger pull.

    Pistols, new or repaired__pounds__ 5½ to 6½
    Pistols, in hands of troops____do____ 5 to 6½

d. Exterior ballistics. (1) *Accuracy with muzzle rest.* The figures indicated below represent the mean variations for several targets.

Range	Mean radi	Mean vertical deviation
Yards	*Inches*	*Inches*
25	0.86	0.62
50	1.36	0.91
75	2.24	1.42

(2) *Drift.* The drift or deviation due to the rifling in this pistol is to the left, but is more than neutralized by the pull of the trigger when the pistol is fired from the right hand. The drift is slight at short ranges and that for long ranges is immaterial, as the pistol is a comparatively short-range weapon.

(3) *Velocity with striking energy.*

Range	Velocity	Energy	Range	Velocity	Energy
Yards	Feet per second	Foot-pounds	Yards	Feet per second	Foot-pounds
0	802	329	150	717	262
25	788	317	175	704	253
50	773	305	200	691	244
75	758	294	225	678	235
100	744	283	250	666	226
125	730	272			

(4) *Penetration in white pine.*

Range	Depth	Range	Depth
Yards	Inches	Yards	Inches
25	6.0	150	5.2
50	5.8	200	4.6
75	5.6	250	4.0
100	5.5		

A penetration of 1 inch in white pine corresponds to a dangerous wound. The penetration in moist loam at 25 yards is about 10 inches. The penetration in dry sand at 25 yards is about 8 inches.

(5) *Trajectory.* The elevation required for 100 yards is 24′ and for 200 yards about 1°. The elevation for 250 yards is about 1°13′; the maximum ordinate being approximately 130 yards distance from the muzzle and about 51 inches in height. The maximum range is approximately 1,600 yards at an angle of elevation of 30°. The maximum ordinate for the maximum range is approximately 2,000 feet.

Section II

DISASSEMBLING AND ASSEMBLING

3. DISASSEMBLING (figs. 3 to 7). **a.** Remove the magazine by pressing the magazine catch.

b. Press the recoil spring plug inward and turn the barrel bushing to the right until the recoil spring plug and the end of the recoil spring protrude from their seat, releasing the tension of the recoil spring. As the recoil spring plug is allowed to protrude from its seat, the finger or thumb should be kept over it so that it will not jump away and be lost or strike the operator. Draw the slide rearward until the smaller rear recess in its lower left edge stands above the projection on the thumbpiece of the slide stop; press gently against the end of the pin of the slide stop which protrudes from the right side of the receiver above the trigger guard and remove the slide stop.

c. This releases the barrel link, allowing the barrel with the barrel link and the slide to be drawn forward together from the receiver, carrying with them the barrel bushing, recoil spring, recoil spring plug, and recoil spring guide.

d. Remove these parts from the slide by withdrawing the recoil spring guide from the rear of the recoil spring and drawing the recoil spring plug and the recoil spring forward from the slide. Turn recoil spring plug to left to remove from recoil spring. Turn the barrel bushing to the left until it may be drawn forward from the slide. This releases the barrel which with the barrel link may be drawn forward from the slide, and by pushing out the barrel link pin the barrel link is released from the barrel.

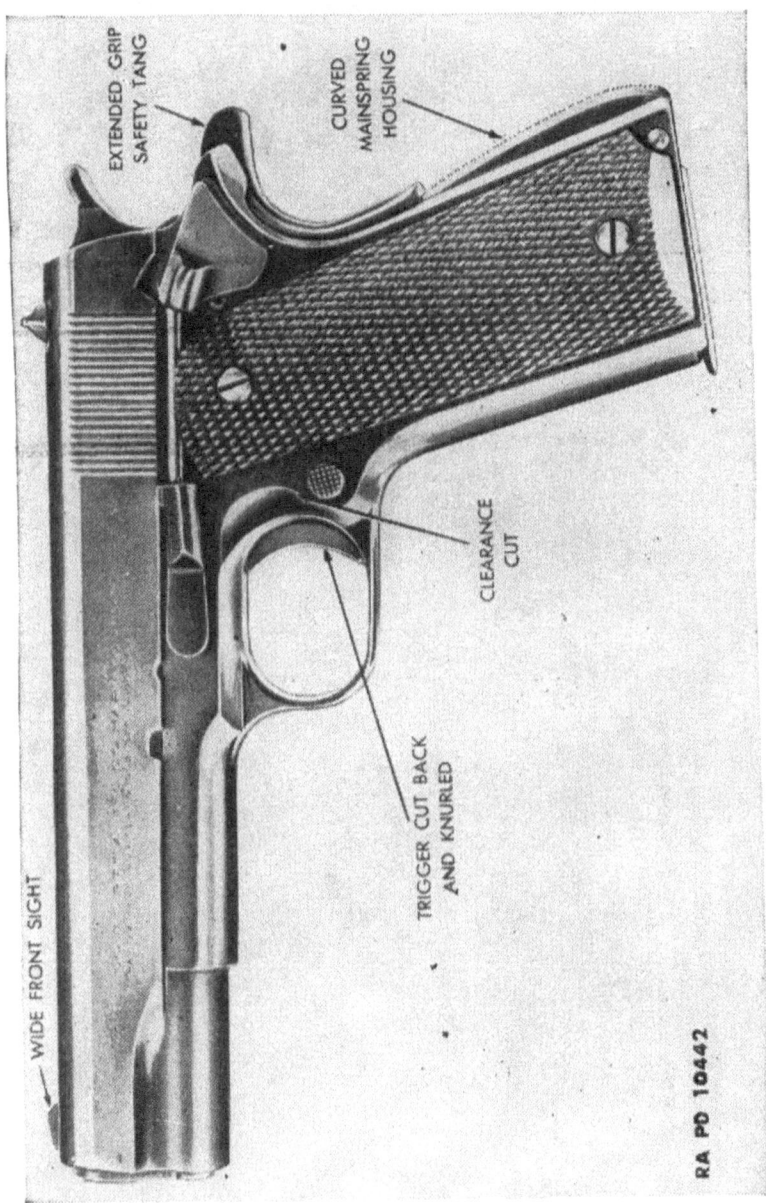

Figure 2. Left side of pistol, M1911A1.

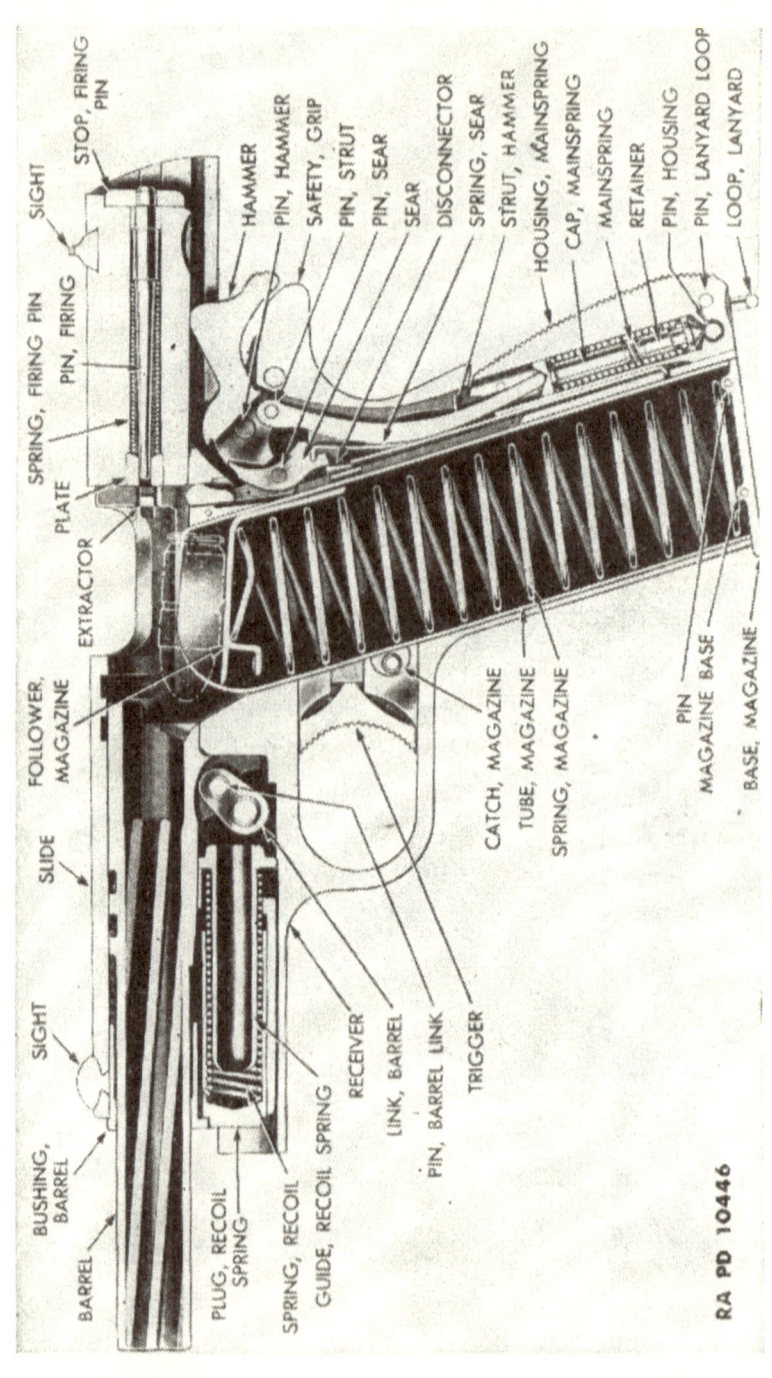

Figure 3. Sectional view of pistol, M1911A1.

e. Press the rear end of the firing pin forward until it clears the firing pin stop which is then drawn downward from its seat in the slide; the firing pin, the firing pin spring, and extractor are then removed from the rear of the slide.

f. The safety lock is readily withdrawn from the receiver by cocking the hammer and pushing from the right on the pin part or pulling outward on the thumbpiece of the safety lock when it is midway between its upper and lower positions. The cocked hammer is then lowered and removed after removing the hammer pin from the left side of the receiver. The mainspring housing pin is then pushed out from the right side of the receiver which allows the mainspring housing to be withdrawn downward and the grip safety rearward from the handle. The sear spring may then be removed. By pushing out the sear pin from the right to the left side of the receiver, the sear and the disconnector are released.

g. To remove the mainspring, mainspring cap, and housing pin retainer from the mainspring housing, compress the mainspring and push out the small mainspring cap pin.

h. To remove the magazine catch from the receiver, its checkered left end must be pressed inward, when the right end of the magazine catch will project so far from the right side of the receiver that it may be rotated one-half turn. This movement will release the magazine-catch lock from its seat in the receiver, when the magazine catch, the magazine catch lock, and the magazine catch spring may be removed.

i. With the improved design of magazine catch lock the operation of dismounting the magazine catch is simplified. When the magazine catch has been pressed inward the magazine catch lock is turned a quarter turn to the left by means of a screw driver, or the short leaf of the sear spring. The magazine catch with its contents can then be removed. The improved design will be recognized from the fact that the head of the magazine catch lock is slotted.

j. The trigger can then be removed rearward from the receiver.

k. The hammer strut or the long arm of the screw driver can be used to push out all the pins except the mainspring cap pin, lanyard loop pin, and ejector pin.

l. The slide stop plunger, the safety lock plunger, and the plunger spring may be pushed to the rear out of the plunger tube.

m. The magazine should not be disassembled except for cleaning or to replace the magazine follower or magazine spring. To disassemble proceed as follows: Push the magazine follower downward about ¼ inch; this compresses the magazine spring. Insert the end of a drift through one of the small holes in the side of the magazine to hold the magazine spring, then slide out the magazine follower. Hold hand over end of the magazine before removing drift from hole to prevent magazine spring from jumping away.

4. ASSEMBLING. a. Proceed in the reverse order.

b. It should be noted that the disconnector and sear are assembled as follows: Place the cylindrical part of the disconnector in its hole in the receiver with the flat face of the lower part of the disconnector resting against the yoke of the trigger. Then place the sear, lugs downward, so that it straddles the dis-

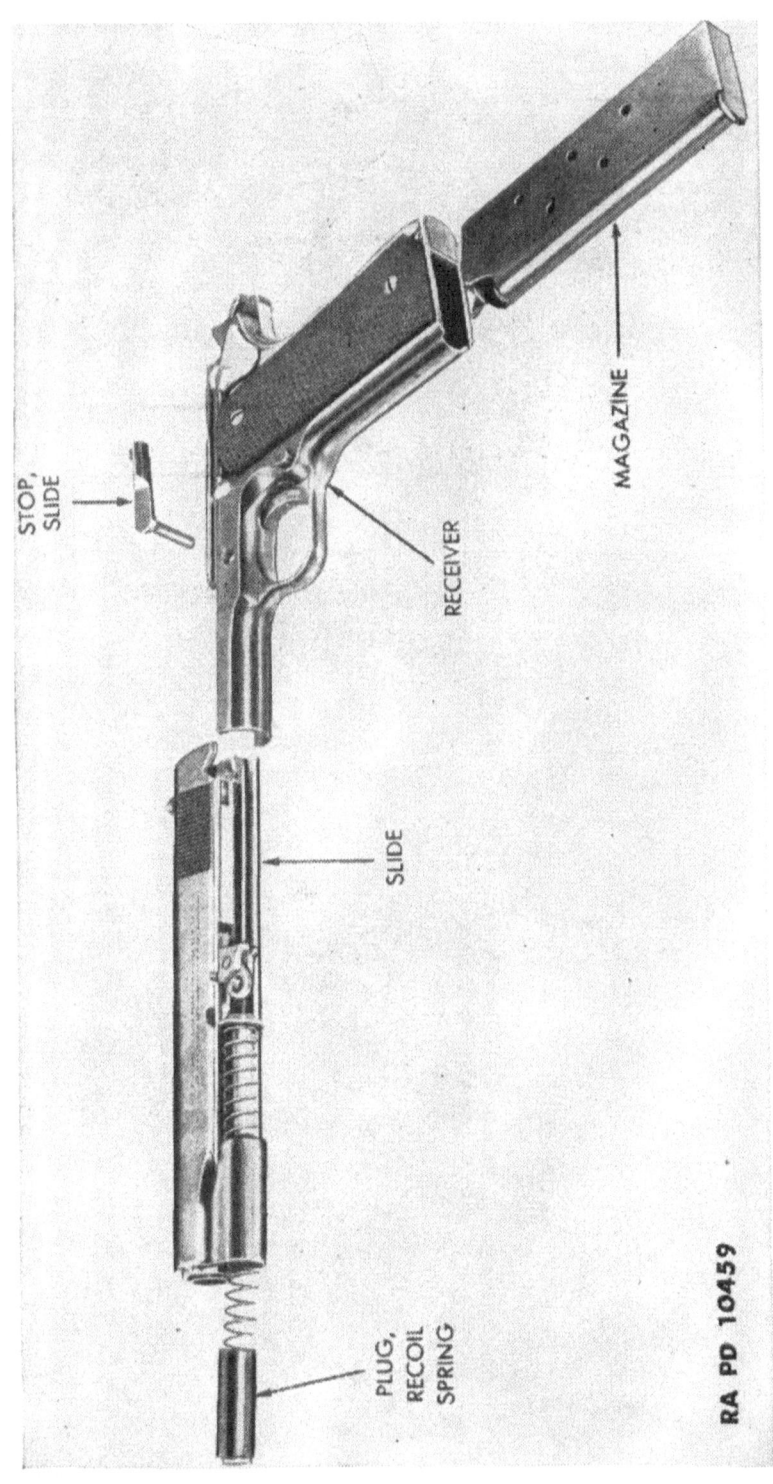

Figure 4. Subassemblies of pistol, M1911A1.

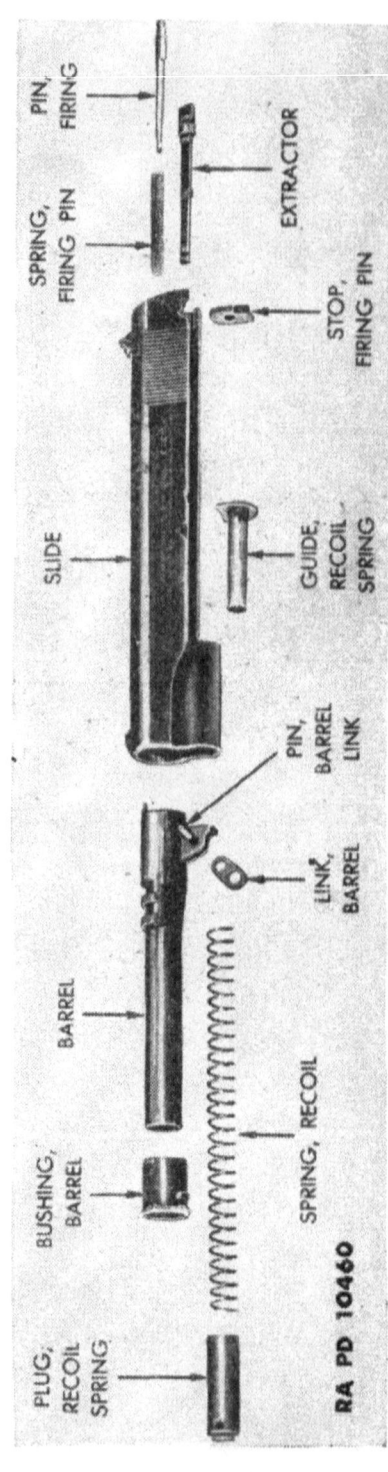

Figure 5. Slide group of pistol, M1911A1.

connector. The sear pin is then inserted in place so that it passes through both the disconnector and the sear.

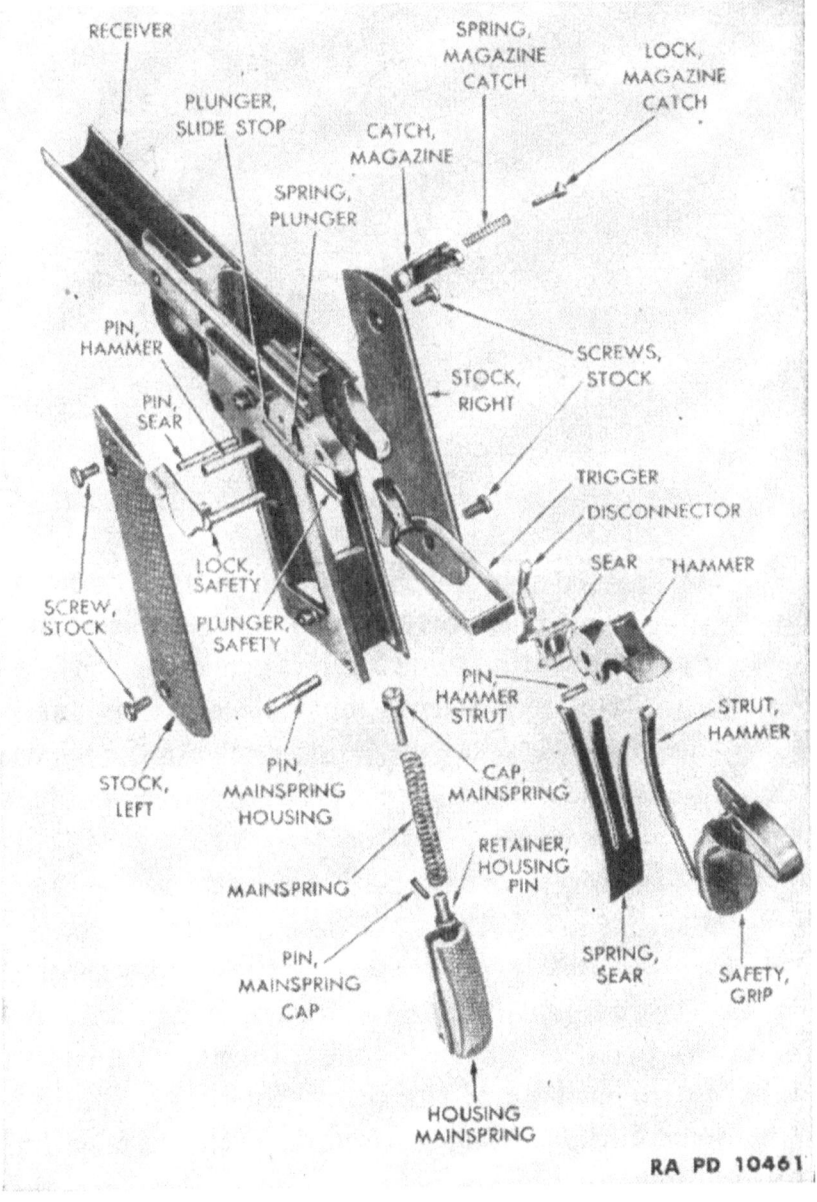

Figure 6. Receiver group of pistol, M1911A1,

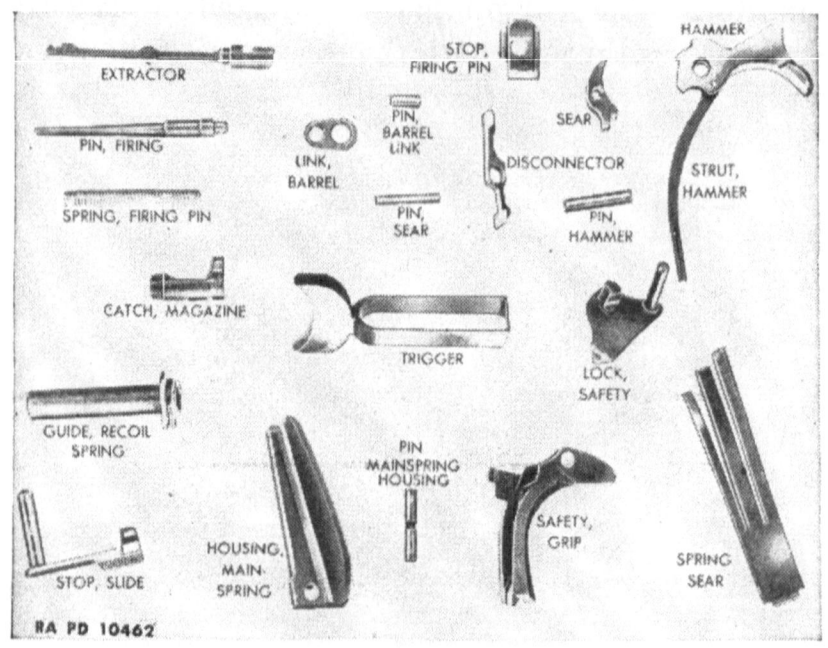

Figure 7. Group of smaller parts of pistol, M1911A1.

c. The sear, disconnector, and hammer being in place and the hammer down, to replace the sear spring, locate its lower end in the cut in the receiver with the end of the long leaf resting on the sear; then insert the mainspring housing until its lower end projects below the frame about ⅛ inch, replace the grip safety, cock the hammer, and replace the safety lock; then lower the cocked hammer, push the mainspring housing home, and insert the mainspring housing pin.

d. In assembling the safety lock to the receiver use the tip of the magazine follower or the screw driver to press the safety lock plunger home, thus allowing the seating of the safety lock. It should be remembered that when assembling the safety lock the hammer must be cocked.

e. When replacing the slide and barrel on the

receiver care must be taken that the barrel link is tilted forward as far as possible and that the barrel link pin is in place.

Section III

CARE AND CLEANING

5. GENERAL. a. Careful, conscientious work is required to keep automatic pistols in a condition that will insure perfect functioning of the mechanism and continued accuracy of the barrel. It is essential that the entire mechanism is kept cleaned and oiled to avoid jams.

b. The mechanism also requires care to prevent rust or an accumulation of sand or dirt in the interior. Pistols are easily disassembled for cleaning and oiling.

6. CARE AND CLEANING. a. Care and cleaning of the pistol include the ordinary care of the pistol to preserve its condition and appearance in garrisons, posts, and camps, and in campaign.

b. Damp air and sweaty hands are great promoters of rust. The pistols should be cleaned and protected after every drill or handling. Special precautions are necessary when the pistols have been used on rainy days and after tours of guard duty.

c. To clean the pistol rub it with a rag which has been lightly oiled, and then clean with a perfectly dry rag. Swab the bore with an oily flannel patch and then with a perfectly dry one. Dust out all crevices with a small, clean brush.

d. Immediately after cleaning, to protect the pistol, swab the bore thoroughly with a flannel patch saturated with oil, lubricating, preservative, special; wipe over all metal parts with an oily rag; apply a few drops of the oil to all cams and working surfaces of the mechanism.

e. After cleaning and protecting the pistol, place it in the pistol rack without any covering whatever. The use of canvas or similar covers is prohibited, as they collect moisture and rust the metal parts. While barracks are being swept, pistol racks will be covered with a piece of canvas to protect the pistols from dust.

7. CARE AND CLEANING AFTER FIRING. a. When a pistol has been fired the bore will be cleaned thoroughly not later than the evening of the day on which it is fired. Thereafter it will be cleaned and oiled each day for at least the next three succeeding days.

b. To clean the bore after firing, first remove slide and barrel. Saturate a patch with rifle bore cleaner. If rifle bore cleaner is not available, dry-cleaning solvent or water may be used. Plain warm water, or even cold water is good, but hot soapy water is better. The cleaning rod with a cloth patch assembled is inserted in the breech and moved forward and back several times. Run cleaning rod, with cleaning brush assembled, back and forth through the bore several times. Again, run several patches saturated with cleaner, solvent, or water through the bore. Follow this with dry patches until they come out clean and dry. Examine bore for cleanliness. If it is not free of all residue, repeat cleaning process. When the bore is clean, saturate a patch in special preservative lubricating oil and run it back and forth through the bore several times.

Caution: After firing do not oil the bore before cleaning.

c. Swab all surfaces of the slide and receiver with a saturated oily patch, followed by dry patches to remove all traces of dust and dirt. Particular attention must be paid to crevices, guides, and guide grooves. When all parts are thoroughly dry and clean, they should be covered with a light coat of oil.

8. RULES FOR CARE OF PISTOL ON THE RANGE. a. Always clean at the end of each day's firing. A pistol that has been fired should not be left overnight without cleaning.

b. Never fire a pistol with any dust, dirt, mud, or snow in the bore.

c. Before loading the pistol make sure that no patch, rag, or other object has been left in the barrel. Such articles collect moisture and are also a hazard if the pistol is fired. Keep chamber free from oil and dirt.

9. CARE AND CLEANING UNDER UNUSUAL CLIMATIC CONDITIONS. a. Cold weather. (1) In temperatures below freezing it is necessary that the moving parts of the weapon be kept free from moisture. Excess oil on working parts will solidify and cause sluggish operation or complete failure.

(2) The weapon should be taken apart and cleaned with dry-cleaning solvent before use in temperatures below 0° F. Working surfaces which show signs of wear may be lubricated by rubbing lightly with a cloth which has been wet with oil, lubricating, preservative, special.

b. Hot weather. (1) In tropical climates where temperature and humidity are high, or where salt air is present, and during rainy seasons the weapon should be inspected daily. It should be kept lightly oiled when not in use. It should be disassembled daily and all parts dried and oiled.

(2) In hot, dry climates where sand and dust may get into the mechanism and bore, all lubricants should be removed from the pistol, and it should be disassembled daily for thorough cleaning. It should be wiped clean as often as required.

(3) Perspiration from the hands is a contributing factor to rust because it contains acid. Metal parts should be wiped frequently.

10. CARE AFTER GAS ATTACKS. a. Pistols should be cleaned as soon as possible after a gas attack.

b. Oil will prevent corrosion for about 12 hours.

c. Clean all parts in boiling water containing a little soda, if available.

d. All traces of gas must be removed from ammunition with a slightly oiled rag; then thoroughly dry the ammunition.

e. If the weapon is actually contaminated with a liquid blister gas, soak up such liquid as soon as possible, using rags or paper.

f. Gasoline, oil, alcohol, or dry-cleaning solvent, if readily available, may be applied to other cloths and used to wipe off the traces of blister gas which remain in the blotted up areas. Do not spread the solvent beyond the area of the spots because a thin film of agent will remain.

g. Lewisite may be destroyed with water. Soapy

water is especially effective. Thorough scrubbing is necessary to remove the blister reaction product formed by the water.

h. To eliminate remaining traces of blister gas, apply DANC or protective ointment. After approximately 15 minutes wipe off the decontaminant with a clean cloth.

i. Coat the decontaminated surface of the weapon (not ammunition) with oil.

j. Bury or burn cloths used in removing blister gas.

11. IMPORTANT POINTS TO BE OBSERVED. a. After firing the pistol, never leave it uncleaned over night. The damage done is then irreparable.

b. Keep the pistol clean and lightly lubricated, but do not let it become gummy with oil.

c. Do not place the pistol on the ground where sand or dirt may enter the bore or mechanism.

d. Do not plug the muzzle of the pistol with a patch or plug. One may forget to remove it before firing, in which case the discharge may bulge or burst the barrel at the muzzle.

e. A pistol kept in a leather holster may rust due to moisture absorbed by the leather from the atmosphere, even though the holster may appear to be perfectly dry. If the holster is wet and the pistol must be carried therein, cover the pistol with a thick coat of oil.

f. The hammer should not be snapped when the pistol is partially disassembled.

g. The trigger should be squeezed with the forefinger. If the trigger is squeezed with the second finger, the forefinger extending along the side of the receiver is apt to press against the projecting pin of the slide stop and cause a malfunction when the slide recoils.

h. Pressure on the trigger must be released sufficiently after each shot to permit the trigger to reengage the sear.

i. To remove cartridges not fired, disengage the magazine slightly and then extract the cartridge in the barrel by drawing back the slide.

j. Care should be taken to see that the magazine is not dented or otherwise damaged.

k. Care must be exercised in inserting the magazine to insure its engaging with the magazine catch. Never insert the magazine and strike it smartly with the hand to force it home, as this may spring the base or the inturning lips at the top. It should be inserted by a quick continuous movement.

Section IV

FUNCTIONING

12. METHOD OF OPERATION. a. A loaded magazine is placed in the receiver and the slide drawn fully back and released, thus bringing the first cartridge into the chamber. (If the slide is open push down the slide stop to let the slide go forward.) The hammer is thus cocked and the pistol is ready for firing.

b. If it is desired to make the pistol ready for instant use and for firing the maximum number of shots with the least possible delay, draw back the slide, insert a cartridge by hand into the chamber of the barrel, allow the slide to close, then lock the slide and the cocked hammer by pressing the safety lock upward and insert a loaded magazine. The slide and hammer being thus positively locked, the pistol may be carried safely at full cock, and it is only necessary to press down the safety lock (which is located within easy reach of the thumb) when raising the pistol to the firing position.

c. The grip safety is provided with an extending horn which not only serves as a guard to prevent the hand of the shooter from slipping upward and being struck or injured by the hammer, but also aids in accurate shooting by keeping the hand in the same position for each shot and, furthermore, permits the lowering of the cocked hammer with one hand by automatically pressing in the grip safety when the hammer is drawn slightly beyond the cocked position. In order to release the hammer, the grip safety must be pressed in before the trigger is squeezed.

13. SAFETY DEVICES. a. It is impossible for the firing pin to discharge or even touch the primer except on receiving the full blow of the hammer.

b. The pistol is provided with two automatic safety devices:

(1) The disconnector, which positively prevents the release of the hammer unless the slide and barrel are in the forward position and safely interlocked. This device also controls the firing and prevents more than one shot from following each squeeze of the trigger.

(2) The grip safety which at all times locks the trigger unless the handle is firmly grasped and the grip safety pressed in.

c. In addition, the pistol is provided with a safety lock by which the closed slide and the cocked hammer can be positively locked in position.

14. DETAILED FUNCTIONING. a. The magazine may be charged with any number of cartridges from one to seven.

b. The charged magazine is inserted in the receiver and the slide drawn once to the rear. This movement cocks the hammer, compresses the recoil spring, and when the slide reaches the rear position the magazine follower raises the upper cartridge into the path of the slide. The slide is then released and being forced forward by the recoil spring carries the first cartridge into the chamber of the barrel. As the slide approaches its forward position, it encounters the rear extension of the barrel and forces the barrel forward; the rear end of the barrel swings upward on the barrel link, turning on the muzzle end as on a fulcrum. When the slide and barrel reach their forward position they are positively locked together by the locking ribs on the barrel and their joint forward movement is arrested by the barrel lug encountering the pin on the slide top. The pistol is then ready for firing.

c. When the hammer is cocked the hammer strut moves downward, compressing the mainspring, and the sear under action of the long leaf of the sear spring engages its nose in the notch on the hammer. In order that the pistol may be fired the following conditions must exist:

(1) The grip safety must be pressed in, leaving the trigger free to move.

(2) The slide must be in its forward position, properly interlocked with the barrel so that the disconnector is held in the recess on the under side of the slide under the action of the sear spring, transmitting in this position any motion of the trigger to the sear.

(3) The safety lock must be down in the unlocked position so that the sear will be unblocked and free to release the hammer and the slide will be free to move back.

d. When the trigger is squeezed the sear is moved, and the released hammer strikes the firing pin, which transmits the blow to the primer of the cartridge. The pressure of the gasses generated in the barrel by the explosion of the powder in the cartridge is exerted in a forward direction against the bullet, driving it through the bore, and in a rearward direction against the face of the slide, driving the latter and the barrel to the rear together. The downward swinging movement of the barrel unlocks it from the slide, and the barrel is then stopped in its lowest position. The slide continues to move to the rear, opening the breech, cocking the hammer, extracting and ejecting the empty shell, and compressing the recoil spring until the slide reaches its rearmost position when another cartridge is raised in front of it and forced into the chamber of the barrel by the return movement of the slide under pressure of the recoil spring.

e. The weight and consequently the inertia of the slide augmented by those of the barrel are so many times greater than the weight and inertia of the bullet that the latter has been given its maximum velocity and has been driven from the muzzle of the barrel before the slide and barrel have recoiled to the point where the barrel commences its unlocking movement. This construction therefore delays the opening of the breech of the barrel until after the bullet has left the muzzle and therefore practically prevents the escape of any of the powder gasses to the rear after the breech has been opened. This factor of safety is further increased by the tension of the recoil spring and mainspring, both of which oppose the rearward movement of the slide.

f. While the comparatively great weight of the slide of the piston insures safety against premature opening of the breech, it also insures operation of the pistol because, at the point of the rearward opening movement where the barrel is unlocked and stopped, the heavy slide has attained a momentum which is sufficient to carry it through its complete opening movement and makes the pistol ready for another shot.

g. When the magazine has been emptied, the pawl-shaped slide stop is raised by the magazine follower under action of the magazine spring into the front recess on the lower left side of the slide, thereby locking the slide in the open position and serving as an indicator to remind the shooter that the empty magazine must be replaced by a loaded one before the firing can be continued. Pressure upon the magazine catch quickly releases the empty magazine from the receiver and permits the insertion of a loaded magazine.

h. To release the slide from the open position, it is only necessary to press upon the thumbpiece of the slide stop, then the slide will go forward to its closed position, carrying a cartridge from the previously inserted magazine into the barrel and making the pistol ready for firing again.

Section V

SPARE PARTS AND ACCESSORIES

15. SPARE PARTS. In time certain parts of the pistol become unserviceable through breakage or wear resulting from continuous usage. For this reason spare parts are provided for replacement purposes. They should be kept clean and lightly oiled to prevent rust. They are divided into two groups, spare parts and basic spare parts.

   a. Spare parts. These are extra parts provided with the pistol for replacement of the parts most likely to fail, for use in making minor repairs, and in general care of the pistol. Sets of spare parts should be kept complete at all times. Whenever a spare part is taken to replace a defective part in the pistol, the defective part should be repaired or a new one substituted in the spare part set as soon as possible. The allowance of these spare parts is prescribed in ORD. 7, SNL B-6.

   b. Basic spare parts. These are sets of parts provided for the use of ordnance maintenance companies and include all parts necessary to repair the pistol. The allowance of basic spare parts is prescribed in ORD. 8, SNL B-6.

16. ACCESSORIES. The names or general characteristics of many of the accessories required with the automatic pistol indicate their use and application. They consist of the holster, lanyard, and pistol cleaning kit, and for post, camp, or station issue, arm lockers and arm racks. The pistol cleaning kit contains cleaning brushes and rods, pistol screw drivers, an oiler, and a small brass can, in which the set of spare parts is carried.

Section VI

AMMUNITION

17. GENERAL. The information in this section pertaining to the ammunition authorized for use in the revolver, caliber .45, M1917, and the pistol, caliber .45, M1911 and M1911A1, includes a description of the cartridges, means of identification, care, and use (fig. 8).

18. CLASSIFICATION. The types of ammunition provided for the pistol and revolver are:

   a. Ball, for use against personnel and light matériel targets.

   b. Blank, for training purposes.

   c. Dummy, for training (cartridges are inert).

   d. Shot, for use to provide small game for food.

   e. Tracer, for use against personnel and light matériel targets and for signaling.

19. LOT NUMBER. When ammunition is manufactured, an ammunition lot number which becomes an essential part of the marking is assigned in accordance with pertinent specifications. This lot number is marked on all packing containers and on the identification card inclosed in each packing box. It is required for all purposes of record, including grading and use, reports on condition, functioning, and accidents in which the ammunition might be involved. Only those lots of grades appropriate for the weapon will be fired. Since it is impractical to mark the ammunition lot number on each individual cartridge, every effort will be made to maintain the ammunition lot number with the cartridges once they are removed from their original packing. Cartridges which have been removed from the original packing and for which the ammunition lot number has been lost are placed in grade 3. It is therefore obvious that when cartridges are removed from their original packings they should be so marked that the ammunition lot number is preserved.

20. GRADE. AR 775-10 provides for the order in which lots and grades of ammunition are to be used. SB 9-AMM4 lists numerically every lot of small-arms ammunition with its correct grade as established by the office of the Chief of Ordnance. Only lots of proper grade will be fired. Grade 3 indicates unserviceable ammunition which will not be fired.

21. IDENTIFICATION. a. Markings. The contents of original boxes are readily identified by the markings on the box. Similar markings on the carton label identify the contents of each carton.

  b. Types and models. (1) One model each of ball, blank, dummy, shot, and tracer cartridges are authorized for use in both the pistol and the revolver. These cartridges are designated:

    (*a*) Cartridge, ball, caliber .45, M1911.
    (*b*) Cartridge, blank, caliber .45, M9.

(*c*) Cartridge, dummy, caliber .45, M1921.
(*d*) Cartridge, shot, caliber .45, M15.
(*e*) Cartridge, tracer, caliber .45, T30.

(2) The blank cartridge is characterized by the tapered case and the absence of a bullet; it can be used only for single shot fire in the pistol. The dummy cartridge is distinguished by its lack of primer and the small hole in the case. The shot cartridge is recognizable by the blunt end, necked case, and absence of a bullet; to extract it the magazine must be removed from the pistol. The tracer cartridge has a red tip on the bullet.

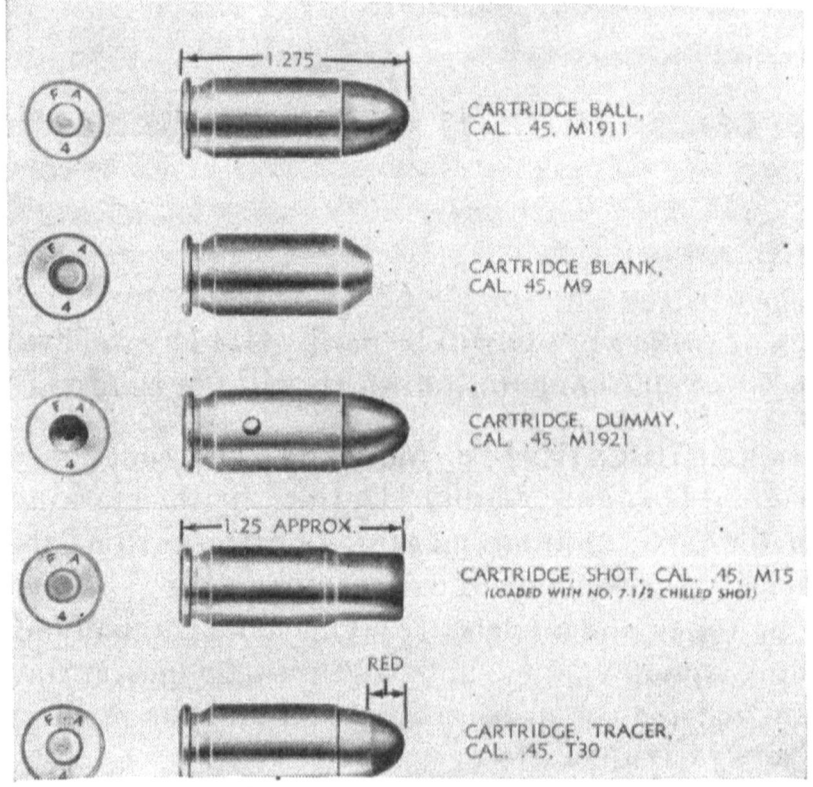

Figure 8. Caliber .45 cartridges.

22. CARE, HANDLING, AND PRESERVATION. a. Small-arms ammunition as compared with other types is not dangerous to handle. However, care must be observed to keep the boxes from becoming broken or damaged.

b. Ammunition boxes should not be opened until the ammunition is required for use. Ammunition removed from the airtight container, particularly in damp climates, is apt to corrode, thereby causing the ammunition to become unserviceable.

c. Carefully protect the ammunition from mud, sand, dirt, and water. If it gets wet or dirty wipe it off at once. If verdigris or light corrosion forms on cartridges it should be wiped off. However, cartridges should not be polished to make them look better or brighter.

d. The use of oil or grease on cartridges is dangerous and is prohibited.

e. Do not fire dented cartridges, cartridges with loose bullets, or otherwise defective rounds.

f. Do not allow the ammunition to be exposed to the direct rays of the sun for any length of time. This is likely to affect seriously its firing qualities.

g. No caliber .45 ammunition will be fired until it has been positively identified by ammunition lot number and grade as published in the latest revision or change to SB 9–AMM 4.

23. STORAGE. a. Whenever practicable small-arms ammunition should be stored under cover. Should it become necessary to leave small-arms ammunition in the open it should be raised on dunnage at least 6 inches from the ground and the pile covered with a double thickness of paulin. Suitable trenches should be dug to prevent water flowing under the pile.

b. Fire hazard. If placed in a fire, small-arms ammunition does not explode violently. There are small individual explosions of each cartridge, the case flying in one direction and the bullet in another. In case of fire it is advisable to keep those not engaged in fighting the fire at least 200 yards from the fire and have them lie on the ground. It is unlikely that the bullets and cases will fly over 200 yards.

Section VII

INDIVIDUAL SAFETY PRECAUTIONS

24. RULES FOR SAFETY. Before ball ammunition is issued, the soldier must know the essential rules for safety with the pistol. The following rules are taught as soon as the recruit is sufficiently familiar with the pistol to understand them. They should be enforced by constant repetition and coaching until their observance becomes the soldier's fixed habit when handling the pistol. When units carrying the pistol are first formed, the officer or noncommissioned officer in charge causes the men to execute INSPECTION PISTOL.

a. Execute UNLOAD every time the pistol is picked up for any purpose. Never trust your memory. Consider every pistol as loaded until you have proved it otherwise.

b. Always unload the pistol if it is to be left where someone else may handle it.

c. Always point the pistol up when snapping it

after examination. Keep the hammer fully down when the pistol is not loaded.

d. Never place the finger within the trigger guard until you intend to fire or to snap for practice.

e. *Never point the pistol at anyone you do not intend to shoot, nor in a direction where an accidental discharge may do harm.* On the range, do not snap for practice while standing back of the firing line.

f. Before loading the pistol, draw back the slide and look through the bore to see that it is free from obstruction.

g. On the range, do not insert a loaded magazine until the time for firing.

h. Never turn around at the firing point while you hold a loaded pistol in your hand, because by so doing you may point it at the man firing alongside of you.

i. On the range, do not load the pistol with a cartridge in the chamber until immediate use is anticipated. If there is any delay, lock the pistol and only unlock it while extending the arm to fire. Do not lower the hammer on a loaded cartridge; the pistol is much safer cocked and locked.

j. In reducing a jam *first remove the magazine.*

k. To remove a cartridge not fired *first remove the magazine* and then extract the cartridge from the chamber by drawing back the slide.

l. In campaign, when early use of the pistol is not foreseen, it should be carried with a fully loaded magazine in the socket, chamber empty, hammer down. When early use of the pistol is probable, it should be carried loaded and locked in the holster or hand. In campaign, extra magazines should be carried fully loaded.

m. When the pistol is carried in the holster loaded, cocked, and locked the butt should be rotated away from the body when drawing the pistol in order to avoid displacing the safety lock.

n. Safety devices should be frequently tested. A safety device is a dangerous device if it does not work when expected.

25. TESTS. a. Safety lock. Cock the hammer and then press the safety lock upward into the safe position. Grasp the stock so that the grip safety is depressed and squeeze the trigger three or four times. If the hammer falls, the safety lock is not safe and must be repaired.

b. Grip safety. Cock the hammer and, being careful not to depress the grip safety, point pistol downward and squeeze the trigger three or four times. If the hammer falls or the grip safety is depressed by its own weight, the grip safety is not safe and must be repaired.

c. Half-cock notch. Draw back the hammer until the sear engages the half-cock notch and squeeze the trigger. If the hammer falls, the hammer or sear must be replaced or repaired. Draw the hammer back nearly to full cock and then let it slip. It should fall only to half cock.

d. Disconnector. Cock the hammer. Shove the slide ¼ inch to the rear; hold slide in that position and squeeze the trigger. Let the slide go forward, maintaining the pressure on the trigger. If the hammer falls, the disconnector is worn on top and must be replaced. Pull the slide all the way to the rear and engage the slide stop. Squeeze the trigger and at the same time release the slide. The hammer should not fall. Release the pressure on the trigger and then squeeze it. The hammer should then fall. The disconnector prevents the release of the hammer unless the slide and barrel are in the forward position safely interlocked. It also prevents more than one shot following each squeeze of the trigger.

CHAPTER 2

MANUAL OF THE PISTOL, DISMOUNTED AND MOUNTED

Section I

GENERAL

26. GENERAL. a. The movements herein described differ in purpose from the manual of arms for the rifle in that they are not designed to be executed in exact unison, there being, with only a few exceptions, no real necessity for their simultaneous execution. They are not therefore planned as a disciplinary drill to be executed in cadence with snap and precision, but merely as simple, quick, and safe methods of handling the pistol.

b. In general, movements begin and end at the position of raise pistol.

c. Commands for firing, when required, are limited to COMMENCE FIRING and CEASE FIRING.

d. Officers and enlisted men armed with the pistol remain at the position of attention during the manual of arms, but render the hand salute at the command PRESENT ARMS, holding the salute until the command ORDER ARMS.

e. When the lanyard is used it should be of such length that the arm may be fully extended without constraint.

Section II

DISMOUNTED

27. RAISE PISTOL (fig.9①). 1. raise, 2. PISTOL. At the command PISTOL, unbutton the flap of the holster with the right hand and grasp the stock, back of the hand outward. Draw the pistol from the holster; reverse it, muzzle up, the thumb and last three fingers holding the stock, the forefinger extended outside the trigger guard, the barrel of the pistol to the rear and inclined to the front at an angle of 30°, the hand as high as, and 6 inches in front of, the point of the right shoulder.

28. WITHDRAW MAGAZINE (fig. 9②). At the command WITHDRAW MAGAZINE, without lowering the right hand, turn the barrel slightly to the right; press the magazine catch with the right thumb and with the left hand remove the magazine. Place between the belt and outer garment.

29. OPEN CHAMBER (fig. 9③). At the command OPEN CHAMBER, withdraw the magazine, if not already withdrawn, and resume the position of raise pistol. Without lowering the right hand, grasp the slide with the thumb and the first two fingers of the left hand (thumb on left side of slide and pointing downward); keeping the muzzle elevated, shift the grip of the right hand so that the right thumb engages the slide stop. Push the slide downward to its full extent and force the slide stop into its notch with the right thumb without lowering the muzzle of the pistol.

① *Raise pistol.*   ② *Withdraw magazine.*

Figure 9. Manual of the pistol.

30. CLOSE CHAMBER. At the command CLOSE CHAMBER, with the right thumb press down the slide stop and let the slide go forward. Squeeze the trigger, being sure that the muzzle is still elevated.

③ *Open chamber.* ④ *Load (pulling slide downward).*

Figure 9—Continued.

31. INSERT MAGAZINE. At the command INSERT MAGAZINE, without lowering the right hand, turn the barrel to the right. Grasp a magazine with the first two fingers and thumb of the left hand; withdraw it from the belt and insert it in the pistol. Press it fully home.

⑤ *Inspection arms.*

Figure 9—Continued.

32. LOAD. At the command LOAD, if a loaded magazine is not already in the pistol, insert one. Without lowering the right hand, turn the barrel slightly to the left. Grasp the slide with the thumb and fingers of the left hand (thumb on right side of slide and pointing upward). Pull the slide downward to its full extent (fig. 9 ④). Release the slide and engage the safety lock.

33. UNLOAD. At the command UNLOAD, withdraw the magazine. Open the chamber as prescribed in paragraph 29. Glance at the chamber to verify that it is empty. Close the chamber. Take the position of raise pistol and squeeze the trigger. Then insert an empty magazine.

34. INSPECTION ARMS (fig. 9⑤). 1. INSPECTION, 2. ARMS. At the command ARMS, withdraw the magazine. Open the chamber as prescribed in paragraph 29. Take the position of raise pistol. The withdrawn magazine is held in the open left hand at the height of the belt. After the pistol has been inspected, or at the command 1. RETURN, 2. PISTOL close the chamber, take the position of raise pistol, and squeeze the trigger, being sure that the muzzle is still elevated. Insert an empty magazine and execute return pistol.

35. RETURN PISTOL. 1. RETURN, 2. PISTOL. At the command PISTOL, lower the pistol to the holster, reversing it, muzzle down, back of the hand to the right; raise the flap of the holster with the right thumb; insert the pistol in the holster and thrust it home; button the flap of the holster with the right hand.

Section III

MOUNTED

36. GENERAL RULES. The following movements are executed as when dismounted: RAISE PISTOL, CLOSE CHAMBER, AND RETURN PISTOL. The mounted movements may be practiced when dismounted by first cautioning, "Mounted position." The right foot is then carried 20 inches to the right and the left hand to the position of the bridle hand. Whenever the pistol is lowered into the bridle hand, the movement is executed by rotating the barrel to the right. Grasp the slide in the full grip of the left hand, thumb extending along the slide, back of the hand down, barrel down and pointing upward and to the left front.

37. WITHDRAW MAGAZINE. At the command WITHDRAW MAGAZINE, lower the pistol into the bridle hand. Press the magazine catch with the forefinger of the right hand, palm of the hand over the base of the magazine to prevent it from springing out; withdraw the magazine and place it between the belt and outer garment.

38. OPEN CHAMBER. At the command OPEN CHAMBER, withdraw the magazine. Grasp the stock with the right hand, back of the hand down, thrust forward and upward with the right hand, and engage the slide stop by pressure of the right thumb.

39. INSERT MAGAZINE. At the command INSERT MAGAZINE, lower the pistol into the bridle hand. Extra magazines should be carried in the belt with the projection on the base pointing to the left. Grasp the magazine with the tip of the right forefinger on the projection, withdraw it from the belt, and insert it in the pistol. Press it fully home.

40. LOAD. At the command LOAD, lower the pistol into the bridle hand. If a loaded magazine is not already in the pistol, insert one. Grasp the stock with the right hand, back of the hand down, and thrust upward and to the left front; release the slide and engage the safety lock.

41. UNLOAD. At the command UNLOAD, withdraw the magazine. Open the chamber as prescribed in paragraph 38. Glance at the chamber to verify that it is empty. Close the chamber. Take the position of raise pistol and squeeze the trigger. Then insert an empty magazine.

42. INSPECTION ARMS. 1. INSPECTION, 2. ARMS. (The pistol is inspected mounted only at mounted guard mounting. The magazine is not withdrawn.) At the command ARMS, take the position of raise pistol. After the pistol has been inspected, or on command, it is returned as prescribed in paragraph 35.

PART TWO
REVOLVER, CALIBER .45

CHAPTER 1

MECHANICAL TRAINING

Section I

DESCRIPTION

43. GENERAL. a. The Colt revolver, caliber .45, M1917, and the Smith and Wesson revolver, caliber .45, M1917, are single shot, breech loading hand weapons. Each is provided with a cylinder having six chambers arranged about a central axis so that six shots may be fired before reloading is necessary. The chambers of the cylinder are loaded with six cartridges in clips of three rounds. When the cylinder is closed, the revolver is ready for firing.

b. These weapons are designed to fire the cartridge, ball, caliber .45, M1911. The action of cocking the

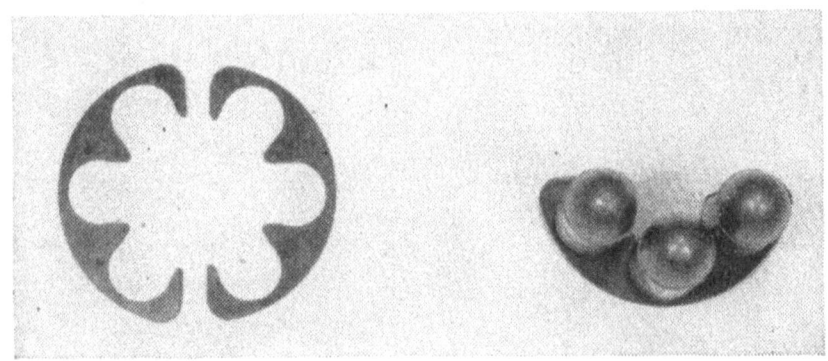

Figure 10. Revolver clip.

hammer causes the cylinder to rotate and align the next chamber with the barrel.

(1) To fire single action, with the revolver fully loaded with ball ammunition, cock the revolver with the thumb and squeeze the trigger for each shot. Double action is executed similarly to single action except that the revolver is cocked by pressing steadily on the trigger.

(2) If one or more of the chambers are empty, the cylinder should be rotated so that a loaded chamber will be moved into line with the barrel when the revolver is cocked. The closed cylinder may be rotated to its proper position by holding the hammer back at about one-fourth full cock. With the hammer of the Colt revolver down, the first loaded chamber should be next on the left of the chamber alined with the barrel, since the cylinder rotates clockwise. With the hammer of the Smith and Wesson revolver down, the first loaded chamber should be next on the right of the chamber aligned with the barrel, since the cylinder of the Smith and Wesson revolver rotates counterclockwise.

c. The rate of fire is limited by the dexterity of the firer in reloading the cylinder and by his ability to aim and squeeze.

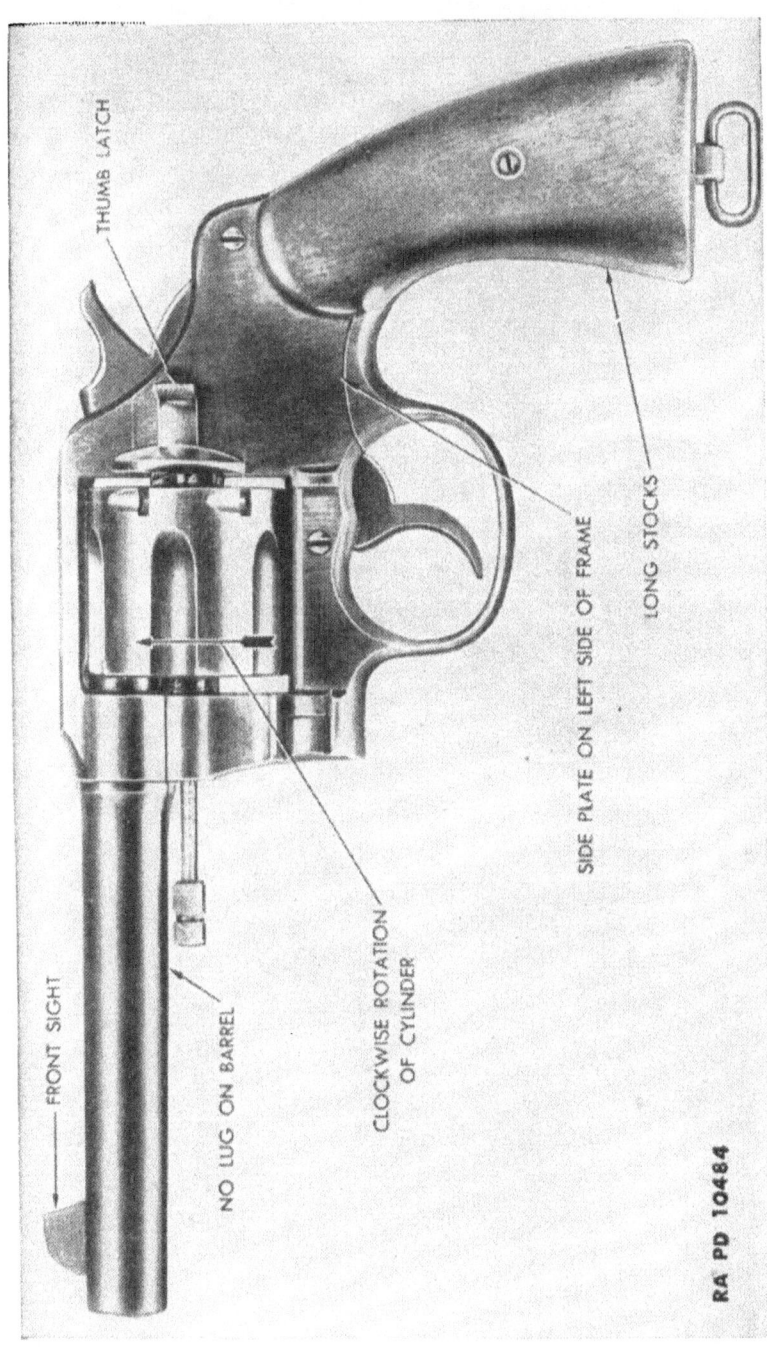

Figure 11. Colt revolver, caliber .45, M1917.

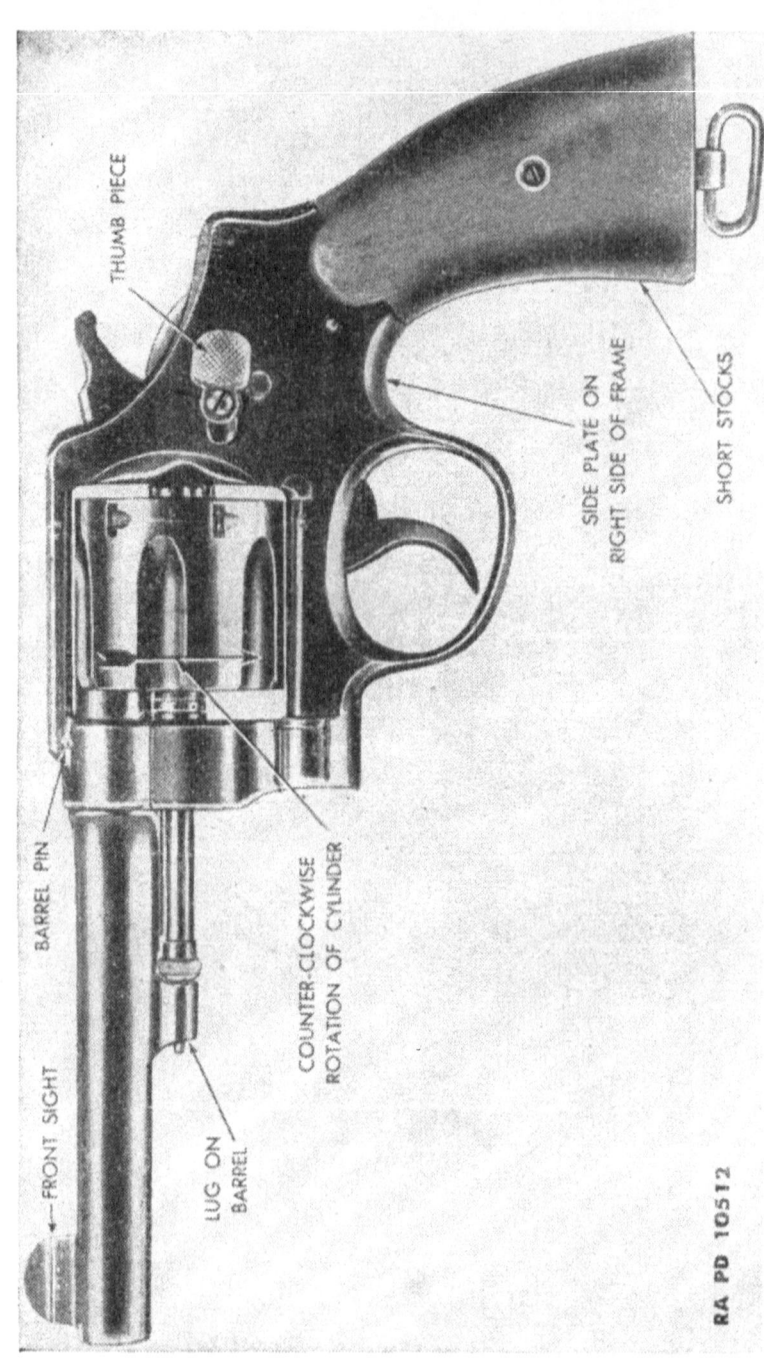

44. TYPES. a. Colt revolver, M1917 (fig. 11).

Weight _____pounds__	2½
Total length_____inches__	10.8
Barrel:	
Length_____do____	5.5
Diameter of bore_____do____	.445
Diameter of rifling_____do____	.452
Rifling, number of grooves_____	6
Grooves:	
Width _____inches__	.156
Depth_____do____	.0035
Twist, one turn in_____do____	16
Lands, width_____do____	.073
Cylinder:	
Length_____do____	1.595
Diameter_____do____	1.695
Chambers:	
Number _____	6
Diameter:	
Maximum_____inches__	.4795
Minimum_____do____	.473
Front sight above axis of bore_do____	.7325

b. Smith and Wesson revolver, M1917 (fig. 12).

Weight _____pounds__	2¼
Total length_____inches__	10.79
Barrel:	
Length_____do____	5.5
Diameter of bore_____do____	.445
Rifling, number of grooves_____	6
Grooves:	
Width _____inches__	.157
Depth_____do____	.003
Twist, one turn in_____do____	14.659

Lands, width_____inches__	.075
Cylinder:	
Length_____do____	1.537
Diameter_____do____	1.708
Chambers:	
Number _____	6
Diameter:	
Maximum_____inches__	.480
Minimum_____do____	.4795
Front sight above axis of bore_do____	.794

Section II

DISASSEMBLING AND ASSEMBLING

45. DISASSEMBLING COLT REVOLVER (figs. 13 to 15). **a.** (1) Remove crane lock screw (19) and crane lock (18) (fig. 13).

(2) Press latch (20) (fig. 14) to the rear, push cylinder to the left, and remove the cylinder and crane assembly by pushing to the front.

(3) Remove stock screw and stocks.

(4) Remove side plate screws.

(5) Remove side plate. Do not pry from its seating. With wooden handle of a tool, tap the plate and frame until the side plate loosens and lift out.

(6) Remove latch and spring from side plate.

(7) Remove mainspring by lifting the rear end from its seat, and disengaging the long end from the hammer stirrup (3) (fig. 15).

(8) Remove the hand (13) (fig. 14).

(9) With a drift, drive the rebound lever pin (38) to the right and remove rebound lever (37) (fig. 15).

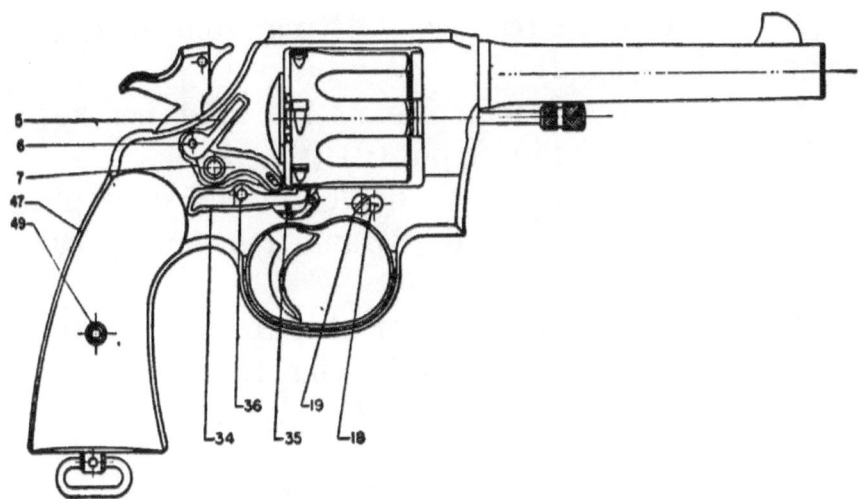

Figure 13. Revolver, Colt, caliber .45, M1917 (right-side view).

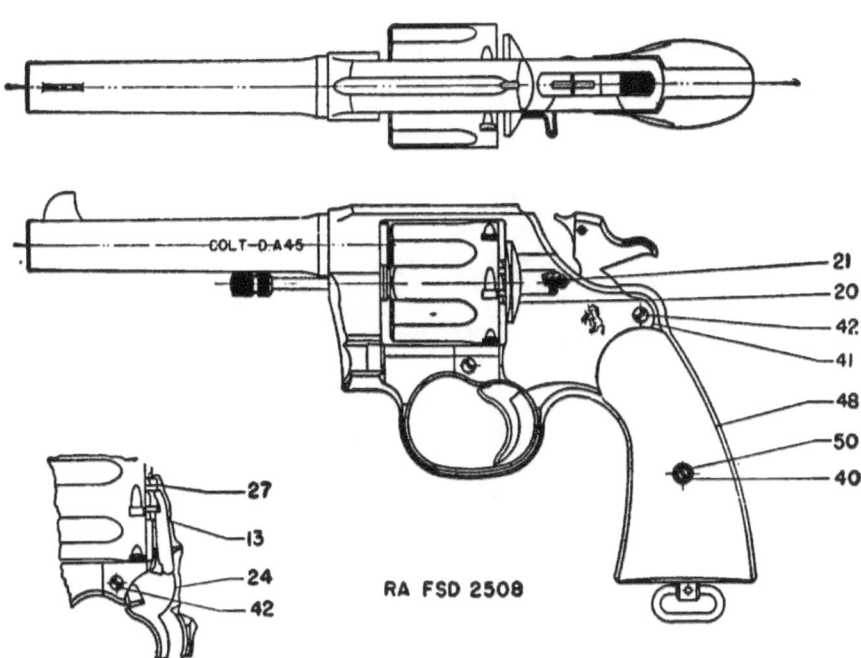

Figure 14. Revolver, Colt, caliber .45, M1917 (left-side and top view).

(10) Remove the trigger by lifting from the trigger pin (25) (fig. 15).

(11) Draw the hammer to its rearmost position and lift from the hammer pin (2) (fig. 15).

(12) With a small drift, drive out the strut pin (10) and remove the strut (8) and strut spring (9) (fig. 15).

(13) Drive out the hammer stirrup pin (4) and remove the hammer stirrup (3) (fig. 15).

(14) Remove safety lever (7) (fig. 13) from its pivot.

(15) Remove safety (5) (fig. 13) from its seat in the frame.

(16) Remove latch bolt (34) (fig. 13) from its seat in the frame.

(17) Remove bolt screw (36) and lift out bolt (34) and bolt spring (35) (fig. 13).

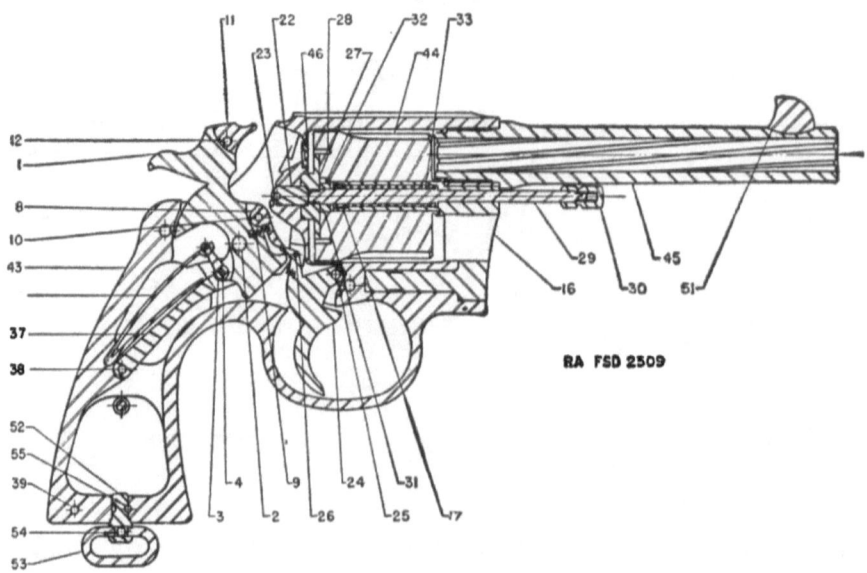

Figure 15. Revolver, Colt, caliber .45, M1917 (sectional view).

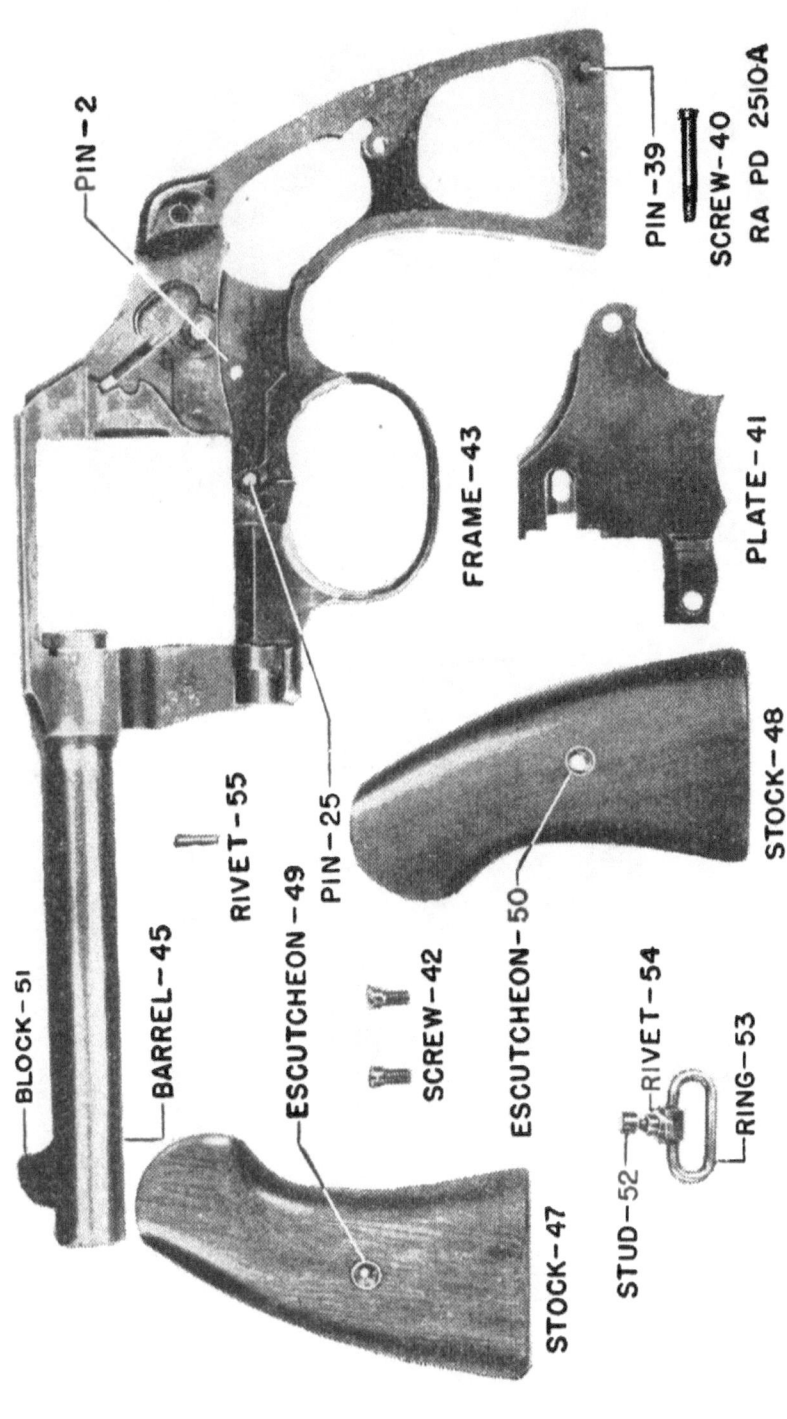

Figure 16. *Revolver, Colt, caliber .45, M1917 parts.*

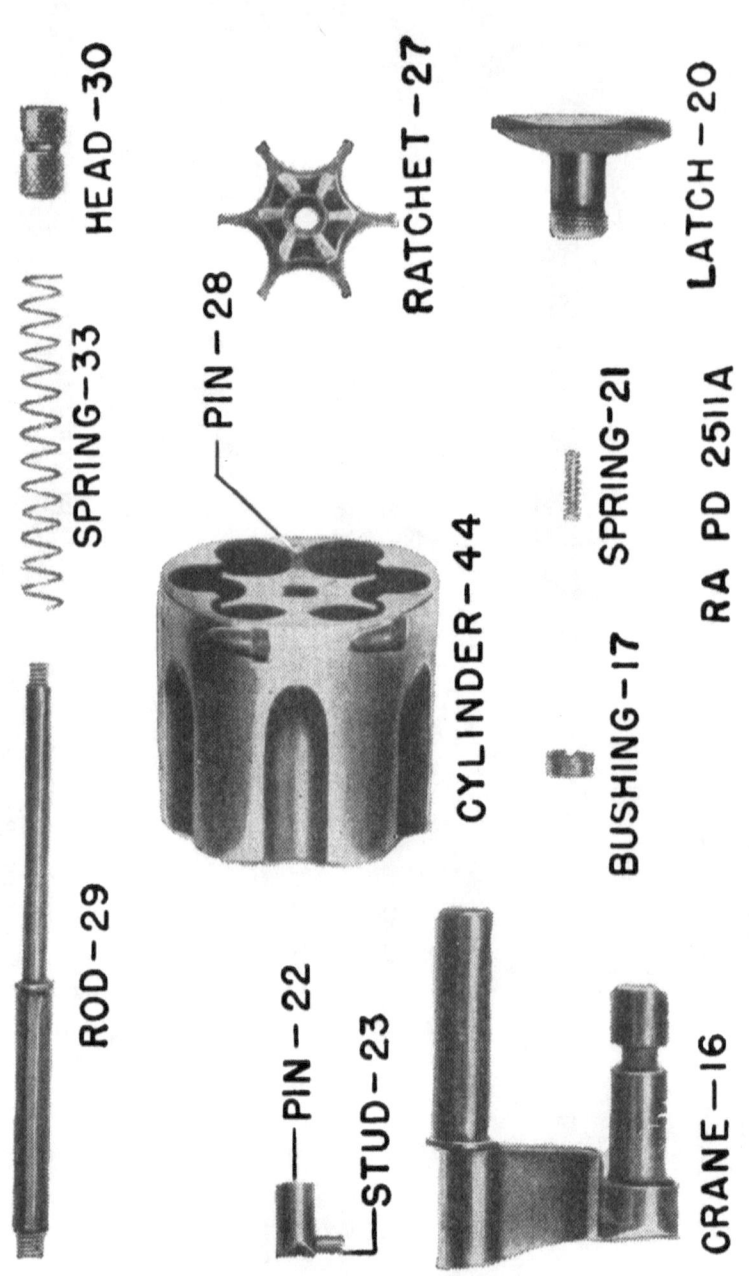

Figure 16. Revolver, Colt, caliber .45 M1917, parts—Continued.

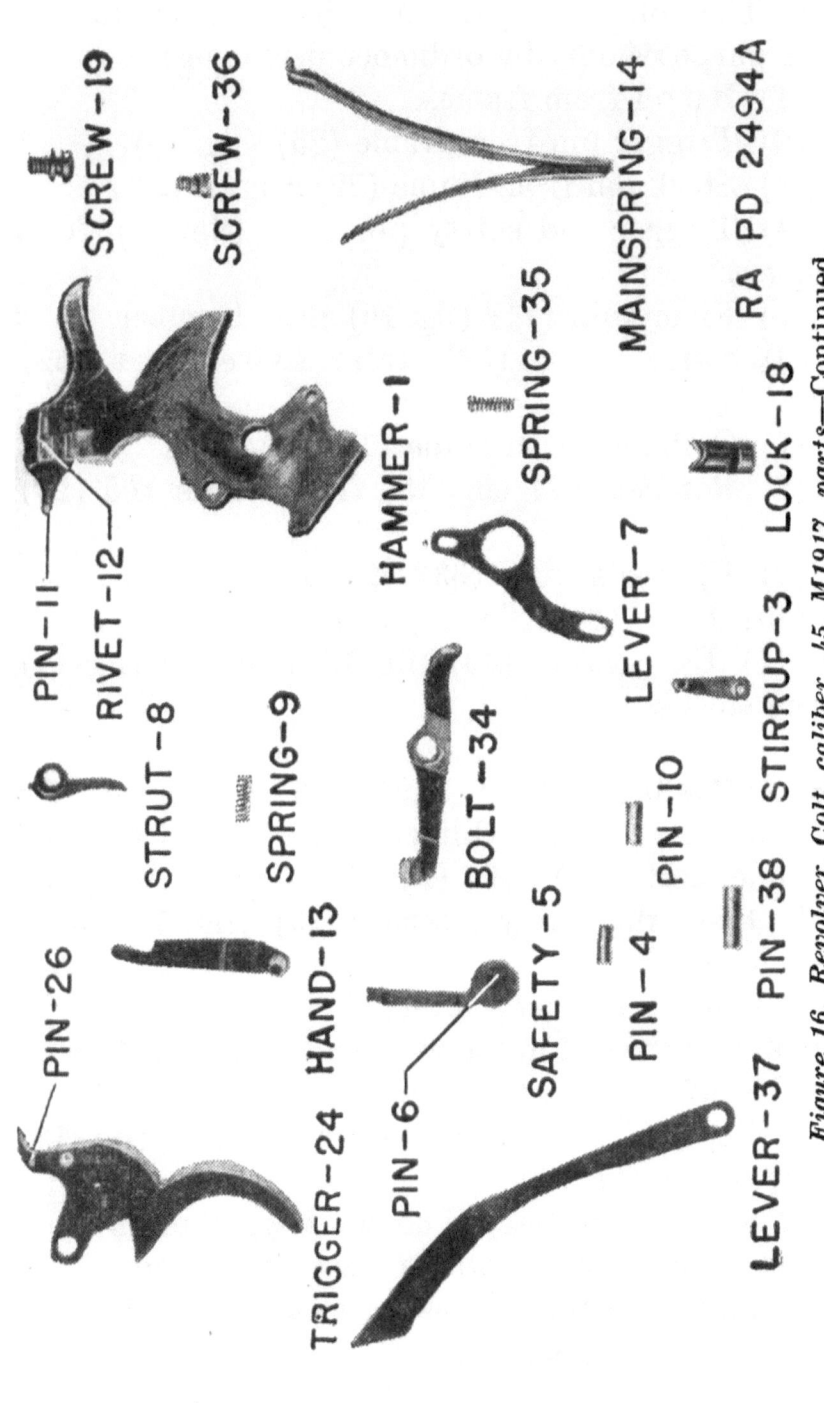

Figure 16. Revolver, Colt, caliber .45, M1917, parts—Continued.

b. The following parts are disassembled for repair purposes only by ordnance personnel:

(1) Barrel from frame.
(2) Trigger pin from frame (25) (fig. 15).
(3) Stock pin from frame (39) (fig. 15).
(4) Trigger and safety pin (26) (fig. 15) from trigger.
(5) Firing pin (11) (fig. 15) from hammer.
(6) Swivel ring (53) from swivel stud (52) (fig. 15).
(7) Cylinder from crane (16) (fig. 15).
(8) Ratchet (27) (fig. 15) from ejector rod (29) (fig. 15).
(9) Ejector spring (33) (fig. 15) from cylinder.
(10) Head (30) (fig. 15) from ejector rod.
(11) Escutcheons (49) (fig. 13) and (50) (fig. 14) from stocks.

46. ASSEMBLING COLT REVOLVER. a. Replace the cylinder bolt (34), cylinder bolt spring (35), and cylinder bolt screw (36) (fig. 13).

b. Place the safety assembly (5) (fig. 13) in its seat in the frame.

c. Place the safety lever (7) (fig. 13) over its pivot with the slot in the short end engaging the stud on the safety.

d. Replace the latch pin assembly (22) (fig. 15) in its seat in the frame.

e. Replace the trigger on the trigger pin so that the stud on the right side of the trigger engages the slot in the longer end of the safety lever (7) (fig. 13).

Note. Test by working trigger forward and back. If the safety lever and safety operate, the assembly is correct.

f. Assemble the strut (8), strut spring (9), and strut pin (10) (fig. 15) to the hammer.

g. Assemble the hammer stirrup (3) and hammer stirrup pin (4) (fig. 15) to the hammer.

h. Place hammer assembly in place on the hammer pin.

i. Assemble the rebound lever (37) to the frame with the rebound lever pin (38) (**fig. 15**).

j. Replace the mainspring so that the notched end engages the hammer stirrup (4) (fig. 15).

k. Insert the stud on the hand (13) (fig. 14) in its hole in the trigger. Press upward on the rebound lever to permit the hand to be fully seated.

l. Replace the latch pin spring (21) (fig. 14) in its seat in the side plate.

m. Put side plate in position but not fully seated.

n. Place the latch in its slot in the side plate so that the latch pin stud (23) (fig. 15) engages in the hole in the latch.

o. Seat the side plate fully and replace the side plate screws.

p. Replace the cylinder and crane, crane lock, and crane lock screw.

q. Replace the stocks and stock screw.

47. DISASSEMBLING SMITH AND WESSON REVOLVER (figs. 17 and 18). **a.** (1) Remove stock screw and stocks.

(2) Remove the side plate screw near the forward part of the trigger guard.

(3) Press forward on the latch to release the cylinder. Push the cylinder to the left and withdraw cylinder and crane assembly to the front, being care-

ful to prevent the crane stop pin (8) and crane stop spring (9) from flying out.

(4) Remove crane stop plunger and spring.

(5) Unscrew thumb piece nut (58) and remove thumb piece (57).

(6) Remove the remaining three side plate screws.

(7) Remove side plate. Do not pry side plate from its seating. With wooden handle of a tool, tap the plate and frame until the side plate loosens, and lift from its seating.

(8) Remove strain screw (56) from recess in butt end of frame.

(9) Remove mainspring (44) by pushing bottom end to the right from its recess in the frame.

(10) Remove rebound slide (45) and rebound slide spring (47) by pressing the rear end of the slide to the right until it clears the rebound slide pin (46).

Note. Hold thumb over rear end of slide as it is removed from the pin in order not to lose the spring.

(11) Remove the hand assembly (33).

(12) Pull the latch (38) back until it clears the rear of the hammer and pull the hammer to the rear. It may be necessary to press the latch away from the frame to allow the hammer to pass. Lift the hammer off the hammer pin (27).

(13) Press trigger assembly to the right and remove from trigger pin.

(14) Remove cylinder bolt plunger screw (15) and cylinder bolt plunger spring (14) and cylinder bolt plunger (13).

(15) Lift cylinder bolt (11) from its pin and remove.

(16) Push latch to rearmost position and remove by pushing the rear end to the right.

(17) Withdraw latch plunger (39) and spring.

b. The following parts are disassembled for repair purposes only by ordnance personnel:

(1) Barrel from frame.
(2) Stock pin (54) from frame.
(3) Rebound slide pin (46) from frame.
(4) Trigger pin (60) from frame.
(5) Cylinder bolt pin (12) from frame.
(6) Firing pin (24) from hammer.
(7) Escutcheons (20 and 21) from stocks.

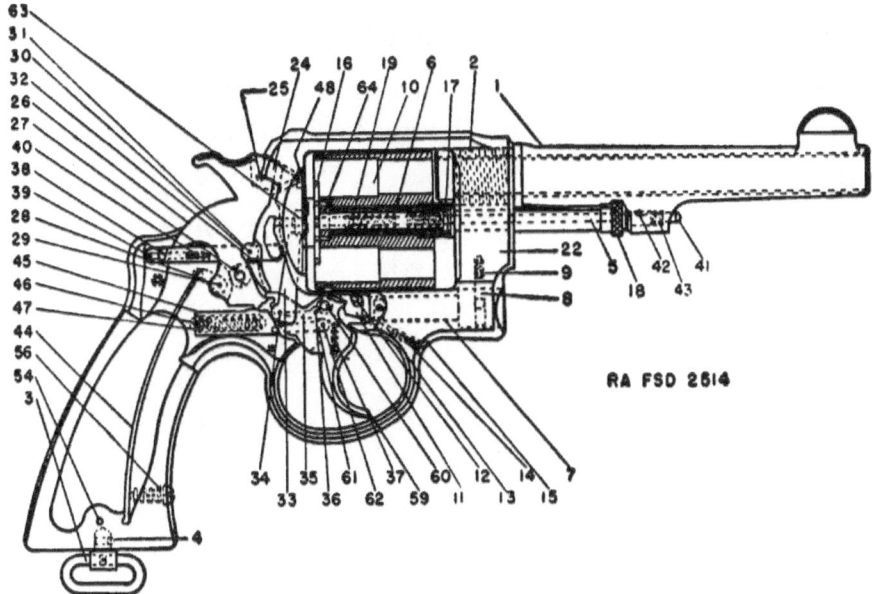

Figure 17. Smith and Wesson revolver, caliber .45, M1917 (sectional view).

48. ASSEMBLING SMITH AND WESSON REVOLVER, M1917 (fig. 17).

a. Replace locking bolt spring (43) and bolt (41) with flat surface up. Replace locking bolt pin (42).

b. Replace cylinder bolt (11) on its pin. Replace cylinder bolt plunger (13), cylinder bolt spring (14), and cylinder bolt screw (15).

c. Assemble the hand (33) to the trigger as follows: With the blade of a screw driver or drift, depress the forward end of the hand lever (35) against the hand lever spring (37). Place the hand pin (34) in its hole in the trigger so that the lug alongside the hand pin is engaged below the rear end of the hand lever.

d. Replace assembled trigger and hand on the trigger pin, holding the upper end of the hand to the rear to clear the frame, and with the rear end of the trigger lever (61) in its topmost position.

e. Replace the latch plunger (39) and latch plunger spring (40) in the recess in the rear end of the latch (38).

f. Replace latch in its guide in the frame by pressing the plunger (39) forward.

g. Replace the hammer assembly on the hammer pin.

Note.—To accomplish this the trigger should be in the rearmost position and the latch should be held to the rear.

h. Put rebound spring (47) into the rebound slide (45) and replace the assembly on the rebound slide pin (46) with beveled end forward, so that the rear end of the trigger lever engages the notch in the forward face of the rebound slide.

i. Replace the mainspring by engaging the hooks on the upper end with the hammer stirrup (28) and then pressing the lower end into its recess in the frame.

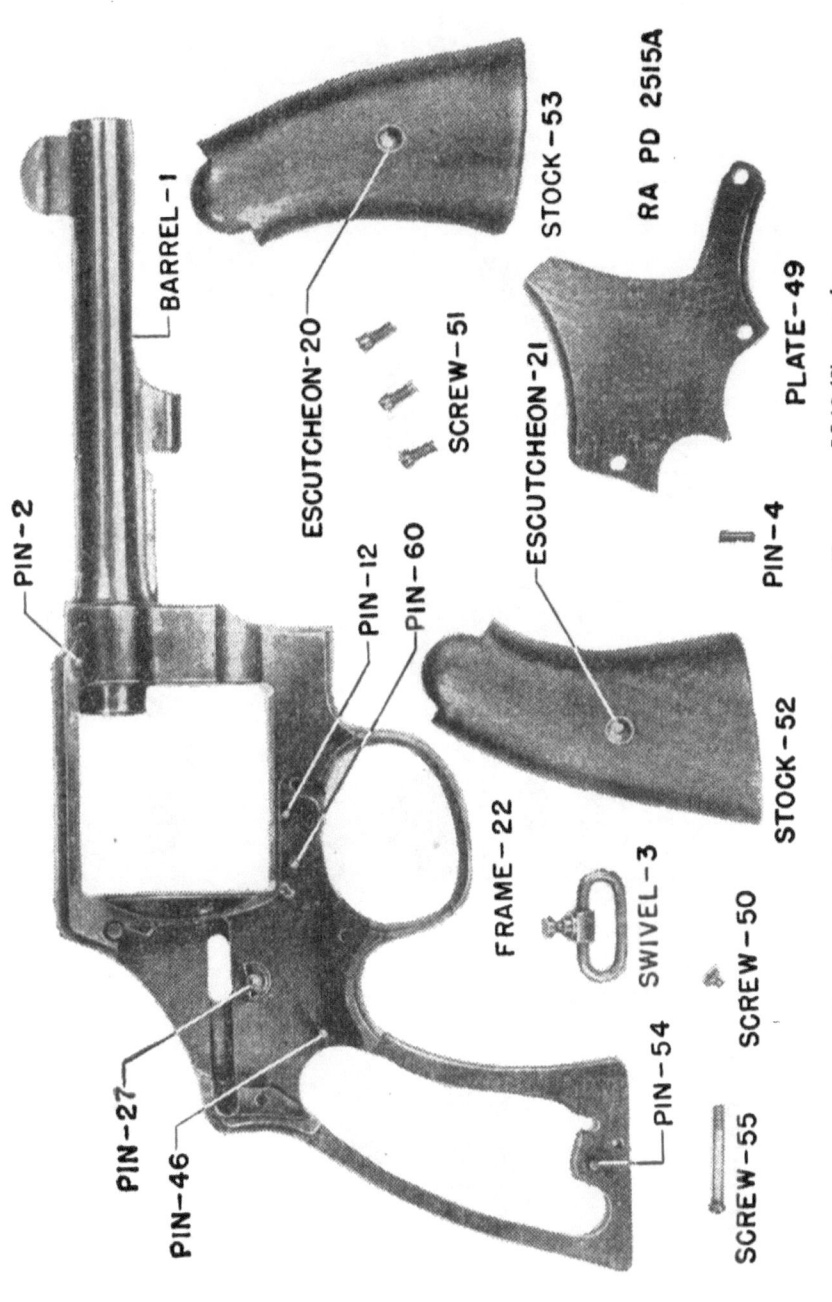

Figure 18. Revolver, Smith and Wesson, M1917, parts.

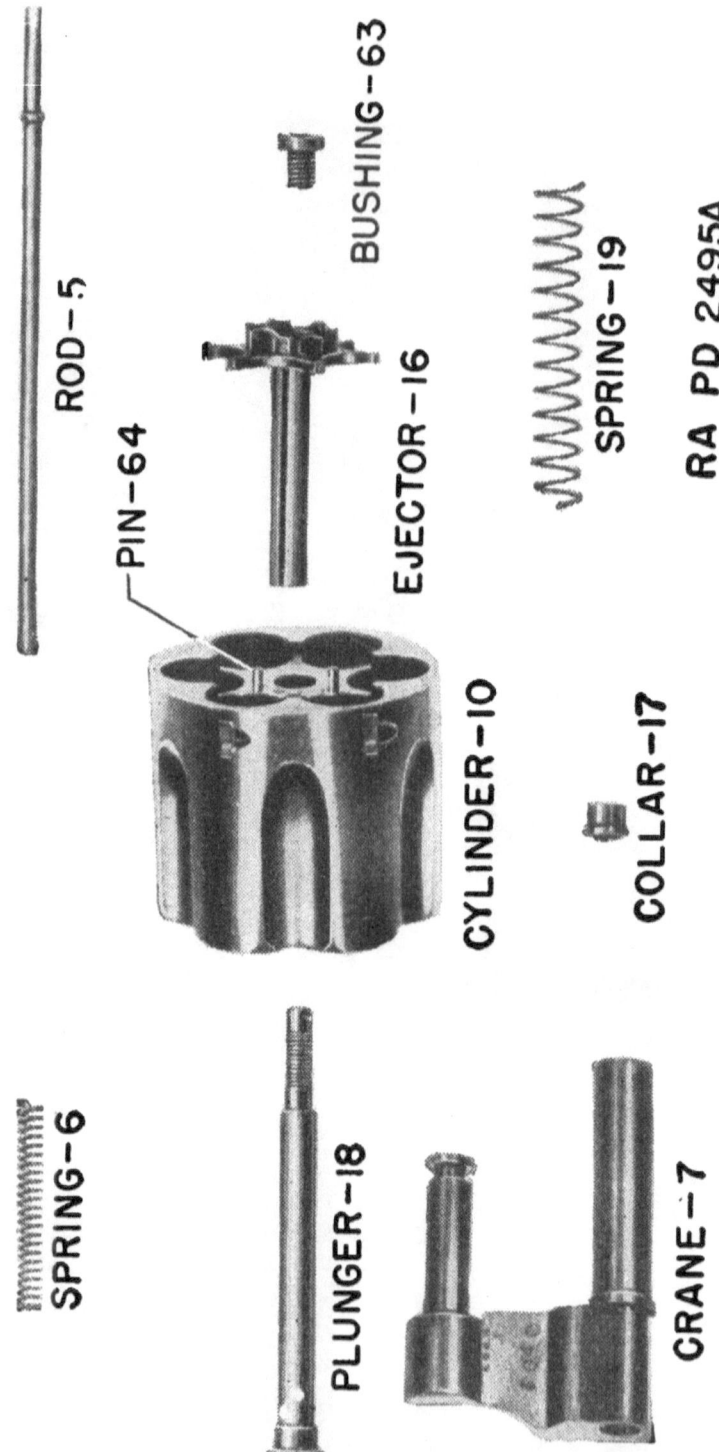

Figure 18. Revolver, Smith and Wesson, M1917, parts—Continued.

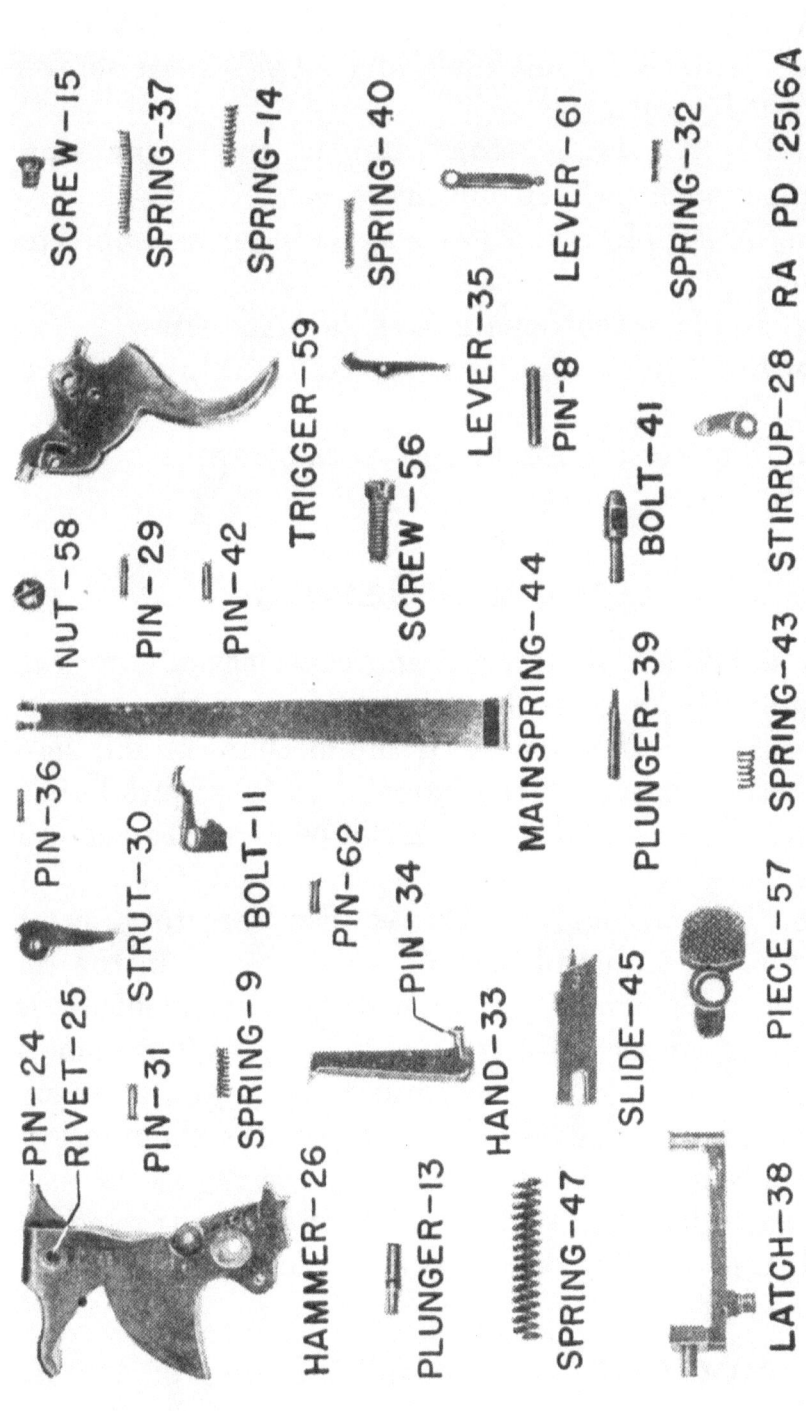

Figure 18. Revolver, Smith and Wesson, M1917, parts— Continued.

j. Replace the mainspring strain screw (56).

k. Replace the side plate and all side plate screws but the forward one.

l. Replace crane stop plunger and crane stop plunger spring (9) in hole in crane.

m. Assemble crane (7) and cylinder assembly to frame.

n. Replace the remaining side plate screw.

o. Replace the thumb piece (57) and thumb piece nut (58).

p. Replace the stocks and stock screw.

Section III

CARE AND CLEANING

49. GENERAL. a. Careful and conscientious work is required to keep revolvers in a condition that will insure perfect functioning of the mechanism and continued accuracy of the barrel. It is essential that exposed parts of the mechanism be kept cleaned and oiled.

b. The mechanism also requires care to prevent rust or an accumulation of sand or dirt in the interior. Revolvers are not usually disassembled for cleaning under ordinary conditions. After immersion in water, after contamination by gas, or if excessive amounts of dirt or sand get into the interior, the side plate should be removed and the mechanism cleaned, dried, and oiled. The side plate is removed only under the supervision of an officer or noncommissioned officer.

50. PROCEDURE. a. Care and cleaning of the revolver include the ordinary care to preserve its condition and appearance in garrisons, posts, and camps, and in campaign.

b. Damp air and sweaty hands are great promoters of rust. The revolver should be cleaned and protected after every drill or handling. Special precautions are necessary when the revolver has been used on rainy days and after tours of guard duty.

c. To clean the revolver, rub it with a rag which has been lightly oiled and then clean with a perfectly dry rag. Swab the bore and chambers with an oily flannel patch and then with a perfectly dry one. Dust out all crevices with a small, clean brush.

d. Immediately after cleaning, to protect the revolver, swab the bore and chambers thoroughly with a flannel patch saturated with oil lubricating, preservative, special; wipe over all metal parts with an oily rag; apply a few drops of oil to all exposed working surfaces of the mechanism. Oil applied in the openings for the trigger, hammer, latch, and cylinder lock will work into the mechanism.

e. After cleaning and protecting the revolver, place it in the revolver rack without any covering whatever. The use of canvas or similar covers is prohibited as they collect moisture and rust the metal parts. While barracks are being swept, revolver racks will be covered with a piece of canvas to protect the revolvers from dust.

51. CARE AFTER FIRING. a. When a revolver has been fired, the bore and chambers will be cleaned thoroughly not later than the evening of the day on which it is fired. Thereafter it will be cleaned and oiled each day for at least the next three succeeding days.

b. To clean the bore after firing, first open the cylinder and hold the revolver with the muzzle pointed downward, toward the operator, and hold the cylinder in its full open position. Saturate a patch with rifle bore cleaner. If rifle bore cleaner is not available, dry cleaning solvent or water may be used. The cleaning rod with a cloth patch assembled is inserted in the muzzle and moved forward and back several times. Run cleaning rod, with cleaning brush assembled, back and forth through the bore several times. Again run several patches saturated with cleaner, solvent, or water through the bore. Follow this with dry patches until they come out clean and dry. Examine bore for cleanliness. If it is not free from all residue, repeat cleaning process. When the bore is clean, saturate a patch in special preservative lubricating oil and run it back and forth through the bore several times.

Caution: After firing do not oil the bore before cleaning.

c. Repeat the operation in b above for each of the chambers of the cylinder, holding the revolver with the muzzle toward the operator, cylinder beneath the frame.

Caution: After firing do not oil the chambers before cleaning.

d. Saturate a clean flannel patch with oil, lubricating, preservative, special, and swab the bore and chambers with the patch, making certain that the bore and all exposed metal parts of the revolver are covered with a light coat of oil.

e. Due to corrosion caused from the gas which

escapes between the barrel and the cylinder, the following parts require special care after firing:

(1) The frame just above the cylinder in the rear of the barrel.

(2) The nose of the hammer.

(3) The firing pin channel and the hammer groove in the frame.

52. RULES FOR CARE OF REVOLVER ON THE RANGE.

a. Always clean at the end of each day's shooting. A revolver that has been fired should not be left overnight without cleaning.

b. Never fire a revolver with any dust, dirt, mud, or snow in the bore.

c. Before loading the revolver make sure that no patch, rag, or other object has been left in the barrel.

d. During range firing, a noncommissioned officer will be placed in charge of the cleaning of revolvers in the cleaning racks.

53. CARE AND CLEANING UNDER UNUSUAL CLIMATIC CONDITIONS.

For care and cleaning of the revolver under unusual climatic conditions, see paragraph 9.

54. CARE AFTER GAS ATTACKS.

For care of revolvers after gas attacks, see paragraph 10.

55. IMPORTANT POINTS TO BE OBSERVED.

a. After firing the revolver, never leave it uncleaned overnight. The damage done is then irreparable.

b. Keep the revolver clean and lightly lubricated but do not let it become gummy with oil.

c. Do not place the revolver on the ground where sand or dirt may enter the bore or mechanism.

d. Do not plug the muzzle of the revolver with a patch or plug. One may forget to remove it before firing, in which case the discharge may bulge or burst the barrel at the muzzle.

e. A revolver kept in a leather holster may rust due to moisture absorbed by the leather from the atmosphere, even though the holster may appear to be perfectly dry. If the holster is wet and the revolver must be carried therein, cover the revolver with a thick coat of oil.

f. The hammer should not be snapped when the revolver is partially disassembled.

g. Pressure on the trigger must be released sufficiently after each shot to permit the trigger to reengage the hammer strut in single action firing, and to permit the trigger to engage the hammer strut in double action firing.

h. The side plate should not be removed except under the supervision of an officer or noncommissioned officer.

i. Never attempt to remove the side plate by prying it out of place.

j. The crane and cylinder of the Colt revolver must not be dismounted except by ordnance personnel.

k. Never attempt to open the cylinder when the hammer is cocked or partly cocked.

l. Never attempt to cock the hammer until the cylinder is fully closed and locked in the frame.

Section IV

FUNCTIONING

56. METHODS OF OPERATION. a. The chambers of the cylinder are loaded with six cartridges, either singly or in clips of three rounds. When the cylinder is closed the revolver is ready for firing.

b. In firing double action, pressure is applied to the trigger until the hammer falls, firing the cartridge.

c. In firing single action, the hammer is cocked by pressure to the rear with the trigger fully released. Pressure on the trigger releases the hammer which falls, firing the cartridge.

d. To lower cocked hammer on a loaded chamber without firing, draw hammer slightly to rear with the thumb; press the trigger to disengage from hammer; let hammer down slowly a short distance, and release trigger. Lower hammer as far as it will go.

57. SAFETY DEVICES. a. General. It is impossible for the firing pin to discharge or even touch the primer except on receiving the full blow of the hammer.

b. Automatic safety devices. (1) *Colt revolver.* The safety lever which is pinned to the trigger moves the safety upward in front of the hammer when the trigger is released after firing a shot. The safety prevents the hammer moving forward sufficiently to strike the primer until pressure is again applied to the trigger, thereby moving the safety downward out of the way. Thus an accidental blow on the hammer cannot cause the revolver to fire. The nose of the cylinder bolt actuated by the cylinder bolt spring projects through a slot in the frame and engages one of the rectangular cuts in the cylinder. This insures positive alignment of one of the chambers of the cylinder with the barrel.

(2) *Smith and Wesson revolver.* A projection on the lower end of the hammer resting against the upper surface of the rebound slide prevents the hammer moving sufficiently far forward to strike the primer, except when the trigger is all the way to the rear. Thus an accidental blow on the hammer cannot cause the revolver to fire. The lug on the upper rear end of the cylinder bolt actuated by cylinder bolt spring projects through a slot in the frame and engages one of the rectangular cuts in the cylinder. This insures positive alignment of one of the chambers of the cylinder with the barrel.

58. DETAILED FUNCTIONING OF COLT REVOLVER, M1917 (fig. 15). **a.** The lock mechanism is contained in the frame and consists of the hammer with its stirrup, stirrup pin, strut, strut pin, and strut spring; the trigger with its pin; the rebound lever; the hand; the cylinder bolt with its spring; the mainspring, which also serves as a rebound lever spring, the hand spring, the trigger spring; the safety and safety lever.

b. The hammer and trigger are pivoted on their respective pins, which are fastened in the right side of the frame. The rebound lever is pivoted on its pin within the grip of the frame. The lower end of the mainspring fits into a slot in the frame, and its upper end engages the hammer stirrup.

c. The lower arm of the mainspring bears on the upper surface of the rebound lever, so that the latter, when the trigger is released after firing a shot, carries the hammer back to its safety position and forces the trigger forward, bringing the hand back to its forward and lowest position. The safety lever, being pinned to the trigger, moves the safety upward in front of the hammer by this same motion.

d. The revolver may be used either single action or double action. In firing double action, pressure upon the trigger causes its upper edge to engage the hammer strut and thereby raises the hammer until nearly in full-cock position, when the strut will escape from the trigger, and the hammer, under action of the mainspring, will fall and strike the cartridge. In firing single action, the hammer is first pulled back with the thumb until the upper edge of the trigger engages in the full-cock notch in the front end of the lower part of the hammer. Pressure on the trigger will release the hammer, which, under the action of the mainspring, will fall and strike the cartridge.

e. The bolt is pivoted on its screw, which is supported in the right side of the frame. The bolt spring pressing upward causes the nose of the bolt to project through a slot in the frame ready to enter one of the rectangular cuts in the surface of the cylinder. During the first part of the movement of the trigger in cocking the revolver, the nose of the bolt is withdrawn from the cylinder by the rear end of the bolt coming into contact with the lug on the rebound lever, permitting the rotation of the cylinder. The object of the bolt is to hold the firing chamber in line with the barrel, and also to prevent the cylinder making more than one-sixth of a revolution at the time of cocking.

f. The hand is attached by its pivot to the trigger, and as the latter swings on its pin when the hammer is being cocked, the hand is raised, revolves the cylinder, and serves with the bolt to lock the cylinder in proper position at time of firing, that is, the axis of the chamber containing the cartridge to be fired coincides with the axis of the bore of the barrel. The pressure of the rebound lever on the lug on the hand insures the engagement of the hand with the ratchet.

g. The cylinder has six chambers. It rotates upon and is supported on the central arbor of the crane. The crane fits into a recess in the frame below the barrel, and turns on its pivot arm, which rotates in a hole in that part of the frame below the opening for the cylinder, and is secured by the crane lock and crane lock screw. The ejector rod passes through the center of the arbor of the crane supporting the cylinder and, projecting under the barrel, terminates in the ejector rod head. The ratchet is screwed on the rear end of the ejector rod with a right-hand thread and then firmly secured by upsetting the end of the rod. The ejector spring is coiled around the ejector rod within the cylinder arbor of the crane, the front end bearing on a shoulder of the rod and the rear end on the crane bushing, which is screwed with a right-hand thread into and closes the cylinder arbor.

h. The latch slides longitudinally on the left side of the side plate, and is connected to the latch pin by the latch pin stud, causing it to follow the movement of the latch. The latch pin slides in a hole in the frame, and when the cylinder is swung into the frame, the latch pin, under action of the latch spring, is forced into a recess in the ejector and locks the cylinder in position for firing. The latch spring is contained in a hole in the side plate in the rear of the latch slot. The recoil plate is driven into its recess in the frame and secured therein by slightly upsetting the rim.

59. DETAILED FUNCTIONING OF SMITH AND WESSON REVOLVER, M1917 (fig. 17). **a.** The lock mechanism is contained in the frame and consists of the hammer, with its stirrup, stirrup pin, strut, strut pin, and strut spring; the trigger, with its pin; the trigger lever, with its pin; the rebound slide, with its pin and spring; the hand, with its pin; the hand lever, with its pin and hand lever spring; the cylinder bolt, with its pin, cylinder bolt plunger, cylinder bolt plunger spring, and cylinder bolt plunger screw; the latch, latch plunger, and latch plunger spring.

b. The hammer and trigger are pivoted on their respective pins. These pins are screwed in place in the left side of the frame and then upset, and are supported on the right side by holes drilled in the side plate to receive them. The rebound slide is held in position by its pin and spring and the rear end of the trigger lever. The lower end of the mainspring fits into a slot in the frame while its upper end engages the hammer stirrup. The mainspring is stressed by screwing up the strain screw, which bears against the mainspring.

c. The rebound slide houses the spring and slides on the pin against which the spring presses. When the trigger is released, after firing a shot, the rebound slide spring pressing against the rebound slide pin and against the end of the recess in the rebound slide forces the rebound slide forward. The forward end of the rebound slide, pressing against the trigger lever, forces the trigger lever forward and returns the trigger to its original position. The hand being pivoted to the trigger by its pin is thus brought back to its lowest position. After firing, when the hammer is in its extreme forward position, the lowest projection on the hammer lies in the notch on the front end of the rebound slide. As the rebound slide moves forward the hammer projection is forced out of the notch and on to the flat surface of the slide in the rear of the notch, thus moving the hammer back to its safety position.

d. The revolver may be used either single action or double action. In firing double action, pressure on the trigger causes its upper edge to engage the hammer strut and raise the hammer until the trigger nose itself actually comes into contact with the hammer. After this, the trigger continues to raise the hammer until the hammer is nearly in its full-cock position, when the hammer will escape from the trigger nose and under action of the mainspring will fall, causing the firing pin to strike the cartridge. In firing single action, the hammer is first pulled back with the thumb until the upper edge of the trigger engages in the full-cock notch in the front end of the lower part of the hammer. Pressure on the trigger will then release the hammer which, under action of the mainspring, will fall and cause the firing pin to strike the cartridge.

e. The cylinder bolt is pivoted on its pin. This is

screwed in the left side of the frame and upset, and on the right side is supported by a hole drilled in the side plate. The cylinder bolt plunger spring pressing upward against the cylinder bolt plunger forces it against the bottom of the cylinder bolt and causes the bolt to project through a slot in the frame ready to enter one of the rectangular cuts in the surface of the cylinder. During the first part of the movement of the trigger in cocking the revolver, the lug on the upper front of the trigger engages in the slot in the cylinder bolt, depresses the bolt, and withdraws the nose of the bolt from the cylinder to permit the rotation thereof. The object of the cylinder bolt is to hold the firing chamber in line with the barrel, and also to prevent the cylinder making more than one-sixth of a revolution at the time of cocking. Accordingly the lug on the trigger releases the cylinder bolt before the hammer falls, thus allowing the bolt to arrest the cylinder.

f. The hand is attached by its pivot to the trigger. As the trigger swings on its pin when the hammer is being cocked, the hand is raised, enters a notch in the ratchet on the ejector, and revolves the cylinder, which is simultaneously freed by the cylinder bolt as above described. It serves with the cylinder bolt, after the bolt is released by the trigger, to maintain the cylinder in proper position at the time of firing, that is, when the axis of the chamber containing the cartridge to be fired coincides with the axis of the bore of the barrel. The hand lever actuated by the hand lever spring, both of which are housed within the trigger, presses against the lug on the hand and insures the engagement of the hand with the ratchet.

g. The cylinder has six chambers. It rotates upon and is supported by the central arbor of the crane. The crane fits into a recess in the frame below the barrel and turns on its pivot arm, which rotates in a hole in that part of the frame below the opening for the cylinder. The ejector plunger enters the arbor of the crane and is held in place by the thread on its rear end which engages the corresponding thread in the front end of the ejector. The shoulder on the ejector plunger bears against the ejector collar so that when the ejector plunger is screwed into its housing in the ejector the ejector spring is compressed. The ejector plunger terminates at its front end in the ejector plunger head. The ejector, of which the ratchet forms a part, consists of a rod and the star-shaped ejector head which engages the clip to cause ejection of the shells. It is forged in one piece. The ejector spring surrounds the ejector within the arbor of the crane, the front end bearing on the ejector collar and the rear end on the shoulder in the rear end of the cylinder.

h. The latch slides longitudinally in its groove within the left side of the frame. It is connected to the thumb piece by the thumb piece nut and is actuated by the latch plunger and latch plunger spring, each housed within the body of the latch. The center rod passes through the ejector plunger and ejector, projecting at the rear end of the cylinder to lock the cylinder in position. The shoulder of the center rod at its rear end bears against the ejector while the ejector plunger bears against the center rod spring, keeping the center rod in place. The front end of the center rod does not come quite flush with the head of the ejector plunger, thus allowing a recess in the head of the ejector plunger into which the locking bolt enters. The latch and the hammer interengage to form an interlock, which prevents cocking of the hammer when the cylinder is unlatched, and prevents unlatching of the cylinder while the hammer is cocked. Since the latch is forced forward by its spring when the cylinder is swung out of the frame, the revolver cannot be cocked unless the cylinder is in position in the frame and latched.

i. The locking bolt is retained in position in its housing by the locking bolt pin and is actuated by the locking bolt spring. The cylinder when closed is retained in its position in the frame and held securely in proper alignment by the center rod which enters its recess in the frame and by the locking bolt, which enters the recess in the head of the ejector plunger. When opening the cylinder, the thumb piece is pushed forward, thus forcing the latch forward. The nose of the latch pushes the center rod out of its recess. The forward movement of the center rod forces the locking bolt forward, thus releasing the cylinder. When the cylinder is swung to the left, out of the frame, it is maintained in its extreme outward position by the crane stop pin, which is forced by the crane stop spring into a small depression drilled in the frame.

i. The frame lug is driven into its recess in the frame and riveted. It serves as a stop to the cylinder when ejection of cartridge cases takes place.

Section V

SPARE PARTS AND ACCESSORIES

60. SPARE PARTS. In time certain parts of the revolver become unserviceable through breakage or wear resulting from continuous usage. For this reason spare parts are provided for replacement purposes. They should be kept clean and lightly oiled to prevent rust. They are divided into two groups: organization spare parts and maintenance spare parts.

   a. Organization spare parts. These are extra parts provided with the revolver for replacement of the parts most likely to fail, for use in making minor repairs, and in general care of the revolver. Sets of spare parts should be kept complete at all times. Whenever a spare part is taken to replace a defective part in the revolver, the defective part should be repaired or a new one substituted in the spare parts set as soon as possible. The allowance of these spare parts is prescribed in ORD 7, SNL B-7.

   b. Maintenance spare parts. These are sets of parts provided for the use of ordnance maintenance companies and include all parts necessary to repair the revolver. The allowance of maintenance spare parts is prescribed in the addendum to ORD 8, SNL B-7.

61. ACCESSORIES. The names or general characteristics of many of the accessories required with the revolver indicate their use and application. They consist of the holster, lanyard, cartridge clips, and pistol cleaning kit; and for post, camp, or station issue, arm lockers and arm racks. The pistol cleaning kit contains cleaning brushes and rods, pistol screw drivers, an oiler, and a small brass can in which the set of spare parts is carried.

Section VI

AMMUNITION

62. REFERENCE. See paragraphs 17 to 23, inclusive.

Section VII

INDIVIDUAL SAFETY PRECAUTIONS

63. RULES FOR SAFETY. Before ball ammunition is issued, the soldier must know the essential rules for safety with the revolver. The following rules are taught as soon as the recruit is sufficiently familiar with the revolver to understand them. They should be enforced by constant repetition and coaching until their observance becomes the soldier's fixed habit when handling the revolver. When units carrying the revolver are first formed, the officer or noncommissioned officer in charge causes the men to execute inspection pistol.

a. Execute unload every time the revolver is picked up for any purpose. Never trust your memory. Consider every revolver as loaded until you have proved it otherwise.

b. Always unload the revolver if it is to be left where someone else may handle it.

c. Always point the revolver up when snapping it after examination. Keep the hammer fully down when the revolver is not loaded.

d. Never place the finger within the trigger guard until you intend to fire or to snap for practice.

e. *Never point the revolver at anyone you do not intend to shoot, nor in a direction where an accidental discharge may do harm.* On the range do not snap for practice while standing back of the firing line.

f. Before loading, open the cylinder and look through the bore to see that it is free from obstruction.

g. On the range do not load the revolver until the time for firing.

h. Never turn around at the firing point while you hold a loaded revolver in your hand, because by so doing you may point it at the man firing alongside of you.

i. On the range do not cock the revolver until immediate use is anticipated. If there is any delay, lower the hammer and recock it only when ready to fire.

j. If the revolver fails to fire, open the cylinder and unload if the hammer is down. If the hammer is cocked or partly cocked, a breakage has occurred. In this case hold the revolver at raise pistol and announce the fact to the officer in charge.

k. To remove a cartridge not fired, open the cylinder and eject, first lowering the hammer if cocked.

l. In campaign the revolver is carried in the holster fully loaded with the hammer down. The cocked revolver should never be put in the holster whether or not it is loaded.

m. The safety device should be tested frequently.

64. TEST OF SAFETY DEVICES. a. Safety. With the revolver unloaded and cylinder closed, cock the hammer. Holding the hammer back with the thumb, press the trigger and let the hammer move forward about ¼ inch, still holding with the thumb. Release the trigger. Then release the hammer and let it fly forward. If the firing pin projects through the hole in the frame, the safety is faulty.

b. Cylinder bolt. With the hammer down attempt to rotate the cylinder. If more than about $\frac{1}{64}$ inch in rotation is possible, the cylinder bolt is faulty. Repeat this test with the hammer fully cocked.

Note. With the hammer about one-fourth cocked the cylinder rotates freely.

CHAPTER 2

MANUAL OF THE REVOLVER, DISMOUNTED AND MOUNTED

Section I

GENERAL

65. GENERAL. a. The movements herein described differ in purpose from the manual of arms for the rifle in that they are not designed to be executed in exact unison, there being, with only a few exceptions, no real necessity for their simultaneous execution. They are not, therefore, planned as a disciplinary drill to be executed in cadence with snap and precision, but merely as simple, quick, and safe methods of handling the revolver.

b. The revolver is used as a substitute for the automatic pistol. For this reason and in the interests of simplicity the term *pistol* is used in all commands.

c. In general, movements begin and end at the position of raise pistol.

d. Commands for firing, when required, are limited to COMMENCE FIRING and CEASE FIRING.

e. Officers and enlisted men armed with the revolver remain at the position of attention during the manual of arms, but render the hand salute at the command PRESENT ARMS, holding the salute until the command ORDER ARMS.

f. When the lanyard is used, it should be of such length that the arm may be fully extended without constraint.

Section II

DISMOUNTED

66. RAISE PISTOL (fig. 19). 1. RAISE, 2. PISTOL. At the command PISTOL, unbutton the flap of the holster with the right hand then turn the back of the hand inward and grasp the stock. Draw the revolver from the holster; reverse it, muzzle up, the thumb and last three fingers holding the stock, the forefinger extended outside the trigger guard, the barrel of the revolver to the rear and inclined to the front at an angle of 30°, the hand as high as, and 6 inches in front of, the point of the right shoulder.

67. LOAD. Being at raise pistol, at the command LOAD, raise the left hand to the front until the forearm is horizontal, palm up. Place the revolver at the cylinder in the left hand, latch up, barrel inclined to the left front and downward at an angle of about

30°. Press the latch with the right thumb, push the cylinder out with the second finger of the left hand and, if necessary, eject the empty shells by pressing the ejector rod head with the left thumb, right hand steadying the revolver at the stock. Take cartridges either singly or in clips from the belt with the right

Figure 19. Raise pistol.

Figure 20. Inspection arms.

hand and insert one in each chamber to be loaded. Close the cylinder with the left thumb and resume the position of raise pistol.

Note. If cartridge clips are not used, empty shells must be removed from the chambers with the fingernails.

68. UNLOAD. Being at raise pistol, at the command UNLOAD, lower the revolver to the left hand and proceed as in paragraph 67, returning unfired cartridges to the belt.

69. INSPECTION ARMS (fig. 20). 1. INSPECTION, 2. ARMS. At the command ARMS, assume the position of raise pistol if not already in that position. Open the cylinder by operating the latch with the right thumb and pushing the cylinder to the left with the right forefinger. After the revolver has been inspected, or at the command **1.** RETURN, 2. PISTOL, close the cylinder with the tip of the right thumb.

70. TO RETURN PISTOL. The commands are: 1. RETURN, 2. PISTOL. At the command PISTOL, lower the revolver to the holster, reversing it, muzzle down, back of the hand to the body; raise the flap of the holster with the right thumb; insert the pistol in the holster and thrust it home; button the flap of the holster with the right hand.

Section III

MOUNTED

71. GENERAL RULES. The following commands are executed as when dismounted: **RAISE PISTOL, LOAD, UNLOAD,** and **RETURN PISTOL.** The mounted movements may be practiced when dismounted by first cautioning, "Mounted Position." At this command, the right foot is carried approximately 20 inches to the right, and the left hand to the position of the bridle hand. Whenever the revolver is lowered into the left hand, the movement is executed by raising the left hand to the front until the forearm is horizontal. Place the revolver in the left hand, latch up, barrel inclined to the left front and downward at an angle of approximately 30°.

72. INSPECTION ARMS. 1. INSPECTION, 2. ARMS. At the command ARMS, lower the revolver into the bridle hand, latch up, barrel inclined to the left front, and downward at an angle of 30°. Press the latch with the right thumb, push the cylinder out with the second finger of the left hand, press the ejector rod head with the left thumb, right hand steadying the revolver at the stock. Bring the revolver up to the position of raise pistol. After the revolver has been inspected, or on command, close the cylinder with the tip of the right thumb, and return revolver to holster as prescribed in paragraph 70.

PART THREE
MARKSMANSHIP

CHAPTER 1

KNOWN-DISTANCE TARGETS

Section I

GENERAL

73. GENERAL. a. Objective. The ultimate objective of training in marksmanship with automatic pistol or revolver is to develop the ability to fire quickly one or more *accurate* shots. To attain accuracy, training must include in its initial phases carefully coached slow fire. *Only after accuracy has been attained* should practice be directed toward development of speed. Extreme care must be exercised to see that speed is not achieved at the expense of accuracy.

b. Scope. Training in marksmanship with hand weapons includes preparatory instruction, range practice and firing for record, and combat firing.

74. METHODS OF INSTRUCTION. a. General. Firing the automatic pistol or revolver is a purely mechanical operation. Therefore, the most effective methods of instruction in marksmanship with hand weapons are the same as those generally applied to instructions in any other mechanical operation. Essential subject material is divided into several progressively arranged phases, each of which the soldier must learn in proper sequence. Further, his work throughout the course of instruction must be supervised with a view to detecting and correcting his mistakes and preventing the fixation of undesirable shooting habits.

b. Coach and pupil method. The coach and pupil method of instruction is peculiarly applicable to training in marksmanship. By working in pairs when receiving instruction, each man of the team is enabled alternately to learn while acting as coach and watching the actions and correcting the mistakes of his partner and then performing the exercises himself as pupil. Each man is also permitted to rest periodically without halting the progress of his training. In order to receive maximum benefit from the coach and pupil method of instruction, each man must understand thoroughly its purpose and his individual responsibility both as coach and pupil. This method of instruction is used throughout all phases of marksmanship training wherever applicable.

c. Individual instruction. (1) Instruction *must be directed toward the individual soldier*. Methods of instruction cannot be successful which neglect the individual and are directed only to a group. The instructor in charge of training and his assistants are, therefore, required to keep themselves informed as to the progress of *each man*. It is their responsibility to ascertain that every soldier understands thoroughly all points of instruction and is capable of applying them. Furthermore, the soldier's knowledge of each principle and its application must be complete before passing on to the next point of instruction. Halfway measures are not acceptable. The individual must perform each exercise correctly before attempting the next.

(2) The closest supervision must be exercised over the performance of all exercises by individuals. Mistakes on their part are noted, brought to their attention, and corrected immediately. This practice is essential to proficient instruction which, in turn, is essential to good marksmanship with hand weapons. As an example, flinching is caused by jerking the trigger. The soldier who jerks the trigger and is not corrected immediately merely continues the malpractice until it has become a fixed habit. Correction of undesirable shooting habits is a long and difficult process when they have become instinctive as a result of poorly supervised practice. However, if the soldier is first thoroughly instructed in the correct principles of firing and is coached properly and with care when he actually begins to shoot, *correct* shooting habits are developed and fixed.

d. Practice. Properly supervised practice is essential to good marksmanship, but practice haphazardly undertaken and inadequately supervised is very likely to cause harm rather than to achieve any benefit. During all practice sessions instructors must impress upon the coaches the absolute necessity for their requiring correct performance by the pupil. Coaches and instructors *must be alert*. They must not allow any individual to persist in erroneous practice which will form an undesirable shooting habit.

e. Instructors. Officers in charge of marksmanship training are responsible that all instructors, commissioned and noncommissioned, are qualified in pertinent subjects. All instructors will adhere to training doctrine enunciated in FM 21-5, TM 21-250, and TF 7-295.

f. Miscellaneous. (1) During all phases of training except record and combat firing, any individual who assumes a firing position must be supervised by a qualified coach. Periods of instruction during the preparatory phases should be conducted out-of-doors with full-sized pistol targets. However, during inclement weather, miniature targets may be used indoors with good results.

(2) Full-size pistol targets should be placed in the vicinity of the barracks to encourage men to spend extra time in practice.

(3) Considerable preparatory exercise is necessary to strengthen the muscles of the hand and the arm and to fix the habit of correct trigger squeeze. Periods of exercise, however, should not ordinarily be of long duration. Three or four 10-minute periods per day for a month will produce good results on the range. Exercises may be conducted concurrently with other training or during periods of inactivity in the field.

(4) Instructors will assure themselves that all weapons used in all periods of training are in satisfactory working condition.

Section II

PREPARATORY TRAINING

75. GENERAL. a. Objective. Preparatory instruction is intended to teach the individual the principles of marksmanship with hand weapons and to assure his proficiency in their application.

b. Scope. Every individual who is to fire on the range must first complete the following phases of preparatory training:

(1) Aiming exercises.
(2) Position exercises.
(3) Trigger-squeeze exercises.
(4) Timed-fire exercises.
(5) Sustained-fire exercises.
(6) Quick-fire exercises.
(7) Examination.

76. CONDUCT OF INSTRUCTION. The phases of instruction indicated in paragraph 75 above, are arranged progressively and are conducted in sequence.

a. Introduction. (1) At the beginning of the first phase of preparatory instruction, the instructor explains the purpose and scope of the entire course of instruction in marksmanship with hand weapons. His explanation should include discussion of the manner in which instruction is organized to keep all men continuously occupied. He further explains fully the coach and pupil method of instruction and its purpose and application. It is advisable that he demonstrate the principles discussed by employment of a demonstration squad.

(2) In addition to the orientation discussion at the beginning of the initial period, each succeeding period of instruction is likewise initiated with a similar discussion relative to that specific period. The instructor explains fully and will demonstrate wherever applicable the subject material and exercises to be covered during the period. He establishes an objective for each man in every period of instruction by explaining fully and clearly exactly what the instruction is intended to achieve.

b. Demonstration. Wherever applicable instructors will employ demonstration squads to emphasize and clarify pertinent items of instruction. Personnel used for demonstration purposes will have been thoroughly rehearsed in their duties and familiarized with the objective and scope of the demonstration they are to perform. Demonstrations will not be haphazardly undertaken, but will be planned in detail and executed in strict adherence to pertinent principles.

c. Application. During applicatory phases of preparatory training the responsibilities incident to supervision by instructors is particularly exacting. The work of coaches as well as pupils must be supervised with great care. Unless sufficient supervision is exercised over them, coaches are apt to be unimpressed with their responsibilities and consequently fail to achieve their purpose. The importance of noting and immediately correcting errors committed by *both* coach and pupil cannot be overemphasized.

77. BLACKENING SIGHTS. In all preparatory firing, except combat firing, both sights of the weapon, if not already sufficiently black, should be blackened to eliminate glare and to silhouette the sights distinctly. Before being blackened the sights should be cleaned and all traces of oil removed. The blackening is done by holding each sight for a few seconds in the point of a small flame which will deposit a uniform coating of lampblack on the metal. A carbide lamp, kerosene lamp, candle, or small pine stick can be used for the purpose. Carbide gas from a lamp is the most satisfactory of the materials named.

78. AIMING EXERCISES. a. Equipment. Equipment should be available in sufficient quantities to provide continuous occupation for all men engaged in training. The instructor will assure himself prior to each period that sufficient equipment is available and is in good condition.

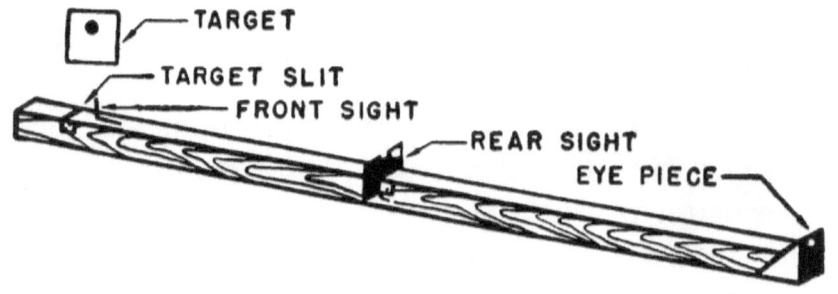

Figure 21. Sighting bar.

(1) *Sighting bar* (fig. 21). (*a*) Pieces of tin or cardboard used on the sighting bar and the top of the bar itself should be blackened carefully. Fasten the bar to a box about 1 foot high and place the box on the ground, a table, or other suitable place.

(*b*) Use of the sighting bar requires the man being instructed to place his eye so that he sees the sights in proper alignment. Since the sights are considerably larger than those on the pistol or revolver, errors are more easily perceived by the instructor and, therefore, more easily pointed out to and understood by the soldier. Use of the eye piece of the sighting bar provides a means by which the instructor may determine if the soldier is placing his eye in the proper position.

(2) *Automatic pistol and revolver rests.* (*a*) To construct a sighting rest for the automatic pistol (fig. 22), use a piece of wood about 10 inches long, 1¼ inches wide, and 9/16 inch thick. Shape one end so that it will fit snugly in the handle of the pistol when the magazine has been removed. Screw or nail this stick to the top of a post or other object at such an angle that the pistol, when placed on the stick, will be aimed approximately parallel with the ground. A suitable sighting rest for the pistol may be improvised easily by cutting an additional notch to hold the pistol in one end of the box used as a rifle rest.

(*b*) To construct a sighting rest for the revolver (fig. 23), build a frame of wood so that the revolver is supported therein. Screw or nail this frame to the top of a post or other object so that the revolver, when in position, is approximately horizontal. In the same manner as for the automatic pistol, a suitable rest for the revolver may be easily improvised by cutting an additional notch in one end of the box used as a rifle rest.

(3) *Aiming disks* (fig. 23). (*a*) For each sighting bar and each pistol rest, a small disk (3 inches in diameter) is made of white cardboard or of tin with white paper pasted on it and with a small bull's-eye in

Figure 22. *The automatic pistol rest and aiming disk in use during sighting exercises.*

Figure 23. The revolver rest and aiming disk in use during sighting exercises.

the center. In the exact center of the bull's-eye is a small hole just large enough to admit the point of a pencil. For indoor or short-range work, the bull's-eye should not be larger than a 50-cent piece.

(b) There should be one 5-inch aiming disk for each squad for shot-group exercises at a range of 25 yards. The large disk should be of tin, painted black, with a handle 4 or 5 feet long and of the same color as the paper on which the shot groups are to be made.

b. First exercise. (1) The instructor displays a sighting bar to his group and points out the front and rear sights, the eyepiece, and the removable target. He explains the following:

(a) The front and rear sights on the sighting bar represent enlarged pistol sights.

(b) The sighting bar is used in the first sighting exercise because, by its use, small errors can be seen easily and explained to the pupil.

(c) Use of the eyepiece requires the pupil to place his eye in such a position that he sees the sights in exactly the same alignment as seen by the coach.

(d) There is no eyepiece on the pistol but the pupil learns by use of the sighting bar how to align the sights properly when using the pistol.

(e) The removable target attached to the end of the sighting bar provides a simple means for readily aligning the sights on a bull's-eye.

(2) The instructor explains the open sight to the assembled group showing each man an illustration of a correct sight alignment (fig. 24).

(3) The instructor, using the open sight, adjusts the sights of the sighting bar with target removed to illustrate a correct alignment of the sights. He re-

Figure 24. Top three sketches illustrate correct sight alignment. Bottom sketch illustrates correct aiming.

quires each man of the assembled group to look through the eyepiece at each of the sight adjustments.

(4) The instructor adjusts the sights to various small errors in sight alignment and requires each man of the assembled group to endeavor to detect the errors.

(5) The instructor describes a correct aim, again showing an illustration to each man (fig. 24). He explains that the top of the front sight is seen through the middle of the open sight in such a manner that its top is level with the outside edges of the open sight and just touches the bottom of the bull's-eye. *All* of the bull's-eye can be clearly seen.

(6) The instructor explains that in aiming, the student's eye should first be focused on the target in order to ascertain that he is aiming at his own target. His eye is then focused on the top of the front sight in order to insure that the line of sight established is a line through the center of the rear sight and over the center of the front sight, whose top is level with the outside edges of the open sight. He also demonstrates how a slight error in centering the front sight in the rear sight will cause the line of sight, when it reaches the target, to diverge greatly from the center of the bull's-eye.

(7) The instructor adjusts the sights of the sighting bar and the removable target to illustrate a correct aim. Each man of the group looks through the eyepiece to observe the correct aiming picture.

(8) The instructor adjusts the sights and the removable target of the sighting bar to illustrate various small errors. Each man in the group attempts to detect and describe them.

(9) After completing the above exercise coaches and pupils are designated, and the exercise is repeated as time allows.

c. Second exercise. (1) With the pistol on the pistol rest and the sights pointing at a blank sheet of paper on a board or wall, the instructor stands with his head in the same relative position as when actually firing the pistol and looks through the sights (fig. 23). By signal or word he has the disk moved until the bottom edge of the bull's-eye is in exact alignment with the sights. Then he commands HOLD and moves away from the pistol. The men then observe the correct sighting and aiming picture through the sights of the pistol.

(2) The instructor requires the man under instruction to look through the sights and direct the disk to be moved until the sights are aligned on the bottom of the bull's-eye. The instructor then looks through the sights to see if the alignment is correct.

(3) The instructor adjusts the sights with the bull's-eye to illustrate various slight errors. He requires the man under instruction to detect them.

d. Third exercise. Using the sighting rest, the man under instruction is required by the instructor to direct the marker to move the disk until the sights are aimed at the bottom edge of the bull's-eye and then to command HOLD. The instructor then checks the aim and, after noticing whether the aim is or is not correct, commands MARK. The marker, without moving the disk, makes a pencil mark on the paper through the hole in the center of the bull's-eye. Repeat the operation until three marks have been made. The instructor checks the aim each time but says nothing to the man until all three marks have been made and joined together. The faults, if any, are then pointed out. The factors contributing to the size and shape of the shot group are discussed and the exercise is repeated several times. At 30 feet, using the small bull's-eye, the shot group should be small enough to be covered by a dime.

79. POSITION EXERCISE. a. Essentials of proper position. To assume the proper position for firing, it is necessary to know the correct position of the body with relation to the target, how to grasp the pistol or revolver, how to aim, and how to hold the breath properly.

(1) *Position of body.* The firer assumes a position on a line perpendicular to the face of the bull's-eye. The body is then faced slightly more than 45° to the left of a line to the bull's-eye. The feet are 12 to 18 inches apart, the head is erect, and the body is in balance when the pistol is held in the firing position. Because of individual peculiarities of conformation, the position may vary slightly, but regardless of bodily conformation, the position assumed should be relaxed and comfortable, and the pistol, when held in the firing position (fig. 25), should point naturally and without undue effort at the center of the target. Unless the body, the pistol, and the target are in correct alignment, muscles of the body must be brought into tension to aim and fire each shot. The muscular tension, in turn, causes trembling and excessive fatigue. If such is the case, the *entire body must be shifted* until the pistol, held in a position to fire, points directly at the center of

Figure 25. Position of the body when firing the pistol. Position is the same with both automatic pistol and revolver.

the target. The position of the body is the same when firing the automatic pistol and the revolver.

(2) *Grasping pistol.* (*a*) With the pistol held in the left hand, the grip safety is forced down and back into the crotch formed between the thumb and forefinger of the right hand. The hand should be as high as possible without having the flesh squeezed between the tang of the hammer and the grip safety. The barrel is aligned with the forearm in such a manner that, if the forefinger were pointed instead of the gun, it would be leveled at the target. The thumb is held parallel to or slightly higher than the forefinger, but should never be lower. The lower three fingers grasp the stock firmly to prevent the recoil from twisting the weapon in the hand and causing the next shot to go wild.

(*b*) The frame of the pistol is squeezed by the thumb and base of the forefinger but, depending upon the conformation of the hand, the *ball* of the thumb may or may not be in contact with the frame. By means of the pressure exerted upon the frame by the thumb and the base of the forefinger, movement of the gun to right or left is controlled and the application of trigger squeeze is more effectively coordinated.

(*c*) The muscles of the arm are firm without being rigid. The barrel is in direct prolongation of the pistol arm and the wrist is locked so that the weapon cannot search up or down. The elbow is straight and locked. The only pivot is at the shoulder joint. After recoil, when the individual is firing correctly, the pistol arm should automatically carry the pistol back to its original position in approximate alignment with the bull's-eye.

Figure 26. Grasping the automatic pistol.

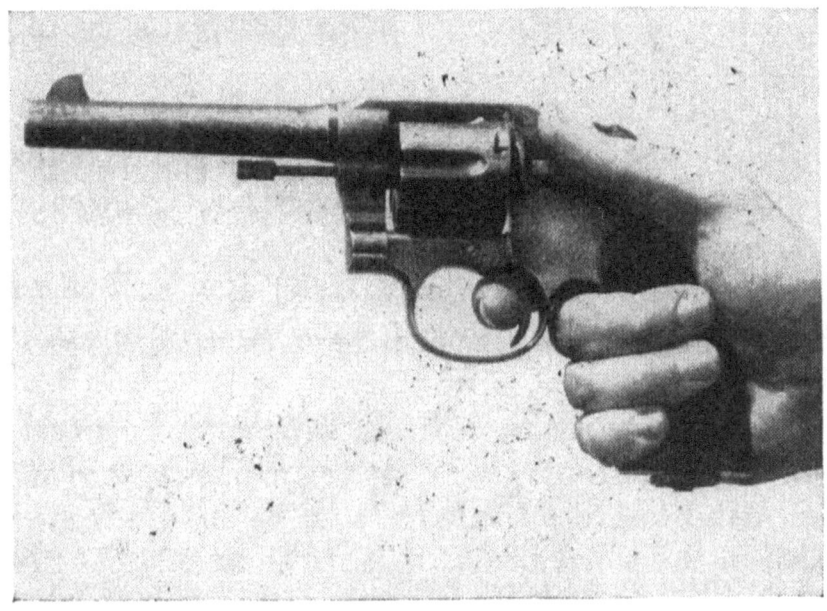

Figure 27. Grasping the revolver.

(*d*) As illustrated in figures 26 and 27, the automatic pistol and revolver are grasped in the same manner except for slight differences caused by conformation of the two weapons.

(3) *Holding breath.* It is important to emphasize that holding the breath properly is necessary to good marksmanship. Emphasis upon this point is required because many men hold their breath improperly or not at all. While the procedure is simple, it requires explanation, demonstration, and supervised practice. To hold the breath properly, an ordinary breath is drawn in, a little let out, and the rest held by closing the throat. The breath should not be held by muscular constriction of the diaphragm since to do so causes muscular strain and trembling and defeats the purpose of holding the breath.

b. Exercise. (1) Required for this exercise: A line of L targets with firing points at 15 and 25 yards or a line of small aiming bull's-eyes placed at the height of the shoulder.

(2) The instructor places the men, armed with the pistol, in one line at 1-pace intervals. He gives the command: 1. INSPECTION, 2. ARMS, and verifies the fact that all pistols are unloaded.

(3) The instructor demonstrates the position of the hand in gripping the pistol and describes the grip in detail.

(4) The instructor requires each man to grip the stock of his pistol in the prescribed manner, using the free hand to grasp the barrel and set the stock well back in the pistol hand between the thumb and the base of the first finger.

(5) The instructor describes the correct method of

holding the breath while aiming. He requires each man to practice the procedure.

(6) The instructor demonstrates the correct position of the body when firing. He explains in detail the position of the feet, legs, body muscles, arms, and head.

(7) The instructor requires each man to assume the correct firing position. Assistant instructors correct individuals who are at fault.

(8) The above exercises having been completed, the men are placed in pairs opposite L targets or opposite small bull's-eyes and are instructed in taking turns coaching each other as follows:

(*a*) Grasp the stock with the correct grip.

(*b*) Face target, then face slightly more than half left.

(*c*) Separate the feet 12 to 18 inches.

(*d*) Align the sights on the bottom edge of the bull's-eye, arm extended.

(*e*) Hold the breath.

(*f*) As soon as the arm becomes tired or the aim becomes unsteady, assume the position of raise pistol.

(*g*) The pistol should be removed from the right hand and the muscles of the hand, arm, and shoulder relaxed and exercised before resuming the grip.

(9) After the pupil has completed the position exercise, he may repeat it using a weight such as a pair of field glasses in a case suspended from the right arm. The weight is suspended first between shoulder and elbow, then from the forearm, then from the wrist and, finally, from the barrel of the pistol. Short rests should be given frequently. The

value of this exercise lies in developing the muscles of the shoulder and arm.

(10) (*a*) The hammer is not raised during the position exercises and the trigger is squeezed very lightly with the finger.

(*b*) After a short rest, repeat the exercise.

(*c*) Coaches watch carefully and correct all errors.

(*d*) The pupil and the coach change places as the instructor desires. This should be every 3 to 5 minutes.

(*e*) Only a few hours should be devoted to the position exercises. All details thereof are included in the trigger-squeeze exercise.

80. TRIGGER SQUEEZE. a. Importance of correct trigger squeeze. (1) The pupil can readily learn to aim and hold the aim either on the bull's-eye or very close to it for at least 10 seconds. When he has learned to squeeze the trigger in such a manner as not to spoil his aim he becomes a good shot. All men flinch in firing the pistol if they know the exact instant at which the discharge is to take place. This is an involuntary action which cannot be controlled. A sudden pressure of the trigger may derange the aim slightly, but the extreme inaccuracy of the shot fired in this way is due mainly to the flinch; that is, the thrusting forward of the hand to meet the shock of recoil. Any man is a good shot who holds the sights of the pistol as nearly on the bull's-eye as possible and continues to squeeze the trigger with a uniformly increasing pressure until the pistol goes off. Any man who has learned to increase the pressure on the trigger only when the sights are in alignment with the bull's-eye, who holds the pressure when the muzzle swerves, and who continues with the pressure when the sights are again in line with the bull's-eye is an *excellent shot*. Any man who tries to "catch his sights" as they touch the bull's-eye and to set the pistol off at that instant is a *very bad shot*.

(2) The apparent unsteadiness of the pistol while being held on the bull's eye does not cause much variation in the striking place of the bullet, because the movement is of the whole extended arm and pistol. However, the sudden pressure of the trigger which always accompanies the flinch deflects the muzzle of the pistol and causes the bullet to strike far from the mark. When the trigger is squeezed, the pressure must be *straight* to the rear. There is a tendency on the part of some men to press the trigger also to the left.

b. Calling shot. To call the shot is to state where the sights were pointed at the instant the hammer fell; for example: "high," "a little low," "to the left," "slightly to the right," or "bull's-eye." If the soldier cannot call his shot correctly in range practice, he has not pressed the trigger properly and consequently does not know where the sights were pointed when the hammer fell.

c. Trigger-squeeze exercises. (1) *First exercise.* (*a*) Required for this exercise: A line of L targets with firing points at 25 yards.

(*b*) The instructor gives the command: 1. INSPECTION, 2. ARMS, and verifies the fact that all pistols are unloaded.

(*c*) The instructor explains and demonstrates the details of this exercise, which are:

1. Cock and lock the piece.
2. Take the correct grip.
3. Take the correct position.
4. Align the sights on the target and start the squeeze, gradually increasing the pressure on the trigger until all the strength of the hand is employed.
5. Rest the hand.
6. Repeat the above operation with the piece unlocked.
7. Call the shot.

(*d*) The instructor assures himself that all the men understand the details of this exercise. The work is then carried on by pairs working together as coach and pupil. The pupil and the coach change places frequently to avoid tiring the muscles of the arm. Extended trigger-squeeze exercise is necessary. The periods utilized for the exercise should be short but frequent.

(*e*) The duties of the coach are—

1. See that the firer uses the correct **grip.**
2. See that the firer assumes a correct position.
3. Watch the hand of the firer to see that he is gradually increasing the pressure on the trigger.
4. See that the firer rests his pistol hand after each time the trigger is squeezed.
5. See that the firer calls the shot when the

exercise is repeated with the piece unlocked.

(2) *Second exercise.* (*a*) The instructor gives the command: 1. INSPECTION, 2. **ARMS**, and verifies the fact that all pistols are unloaded.

(*b*) The instructor explains and demonstrates the details of this exercise, which are:

1. Cock the piece.
2. Take the correct grip.
3. Take the correct position.
4. Align the sights on the target and begin the squeeze. Close the eyes and continue to the squeeze until the hammer falls.
5. When the hammer falls, open the eyes and check the aim to see if it has been deranged.

Note. The firer should be able to keep on the target. If he is persistently off, he should check on his grip and position to see that they are correct.

(*c*) The duties of the coach in this exercise are the same as in the first trigger-squeeze exercise.

(3) *Third exercise.* In learning to fire the pistol, the average man has a tendency to push or punch forward with the arm or shoulder to meet the force of recoil of the piece. To assist the firer in overcoming this tendency, the following exercise is prescribed:

(*a*) By command, the instructor verifies the fact that all pistols are unloaded.

(*b*) The instructor requires the coaches and pupils to take positions at the firing points.

(*c*) The instructor requires the pupil to cock and

lock the piece, take his firing position, align the sights under the bull's-eye, and squeeze the trigger.

(*d*) The coach stands in front of the firer, facing him, and strikes the muzzle of the piece with the palm of his hand. At irregular intervals, his hand misses the muzzle. The firer should hold the piece on the target and make no forward punching movement to meet the shock of the blow. If the firer does push forward with the arm or shoulder, it will be apparent when the coach misses the muzzle.

81. TIMED FIRE. a. Training for timed fire. (1) Training for timed fire is begun after the principles of slow fire, particularly the trigger-squeeze exercise, are thoroughly understood and some facility in applying them has been gained. The trigger-squeeze exercise in slow fire, however, should be resumed and continued during the entire period of preparatory training.

(2) Timed fire is the same as slow fire except that the time between shots is limited. The restriction placed upon the time makes the step from slow fire, without any time limit, to sustained fire, with only a few seconds allowance, an easier one.

(3) The first shot should be fired without undue delay. Succeeding shots should be approximately evenly spaced in order that the last shot will be fired immediately prior to the expiration of the time limit. Although the grasp on the pistol (revolver) may be relaxed after each shot, the proper position and grasp should be taken for every shot. Also for each shot, the breath should be held, the sights correctly aligned, and the trigger properly squeezed.

(4) Immediately after each shot the piece should be cocked, using either hand.

b. Timed-fire exercise. (1) Required for this exercise: A row of L targets or a row of bull's-eyes.

(2) The instructor gives the command: 1. INSPECTION, 2. ARMS, and verifies that all pistols are unloaded.

(3) The instructor explains to the group that the trigger squeeze is the same in timed fire as in slow fire.

(4) Then the instructor takes the proper position and grasp of the piece to simulate fire on a target. He proceeds to explain and demonstrate the correct method of simulating the firing of five shots in the prescribed length of time. The demonstration should include the recocking of the pistol (revolver) immediately after each shot.

(5) (*a*) Upon the completion of the above demonstration the men are placed in pairs in front of the line of targets, one to practice and one to coach. The exercise is then conducted to simulate timed fire in range practice. If a line of disappearing targets is being used for the exercise, the targets appear, remain in sight the allotted time, and then disappear. While the targets are in sight each man undergoing instruction attempts to fire five shots (simulated fire), cocking the piece for each shot, except the first.

(*b*) If the targets are stationary, the exercise begins with the command: 1. COMMENCE, 2. FIRING, and ends with the command: 1. CEASE, 2. FIRING.

(*c*) After each three or four scores of simulated fire, the coaches and pupils exchange duties.

(*d*) In this exercise, the coach notes and corrects all errors in grip, position, trigger squeeze, cocking the piece, and holding the breath. Sustained-fire exercises should be frequent but not repeated until they become tiresome.

82. SUSTAINED FIRE; AUTOMATIC PISTOL. a. Training for sustained fire.

(1) Training for sustained fire is begun after the principles of slow fire and timed fire are thoroughly understood and some facility in applying them has been gained.

(2) The time consumed in squeezing the trigger must necessarily be shorter in sustained fire than in slow or timed fire, but the process is the same.

(3) To fire the first shot, the pistol is brought by the shortest route from raise pistol to the aiming position and sights are aligned on the mark. This is done by smoothly and rapidly extending the right arm straight from the shoulder, inserting the right forefinger in the trigger guard during the movement, and holding the breath. To bring the pistol through the arc of a circle to the aiming position is an unnecessary loss of valuable time.

(4) For succeeding shots, the sights should be held as nearly on the mark as possible and the breath held throughout the score. The recoil after each shot will throw the sights out of alignment, but they should be brought back immediately to the mark by the shortest route. An upward movement of the arm, also caused by recoil, moves the hand 6 to 8 inches. There should be no snapping or bending of the wrist or elbow. The sights will then come back on the mark automatically after each shot. It is a useless

loss of time to give the pistol a flourish between shots.

(5) To simulate the self-loading action of the automatic pistol, the end of a strong cord about 4 feet long is tied to the thumbpiece of the hammer, *the knot on top*, and a few turns of the other end of the cord are taken around the thumb or fingers of the left hand. The cord should be long enough to permit the left hand to hang naturally at the side while aiming the pistol with the right hand, right arm fully extended.

(6) Each time the hammer falls, a quick, backward jerk of the left hand recocks the pistol and, at the same time, jerks the sights out of alignment with the bull's-eye. This derangement of the alignment corresponds very closely to the jump of the pistol when actually firing.

(7) If the knot is underneath the hammer or if a very thick cord is used, the hammer will not remain cocked when jerked back.

b. Sustained-fire exercise. (1) Required for this exercise: A piece of strong cord about 4 feet long for each man; a row of L targets or a row of bull's-eyes.

(2) The instructor gives the command: 1. INSPECTION, 2. ARMS, and verifies the fact that all pistols are unloaded.

(3) The instructor explains to the group that the trigger-squeeze is the same in sustained fire as in slow or timed fire.

(4) Then, the instructor demonstrates the correct method of bringing the pistol to the aiming position by the shortest route. He shows how this is done

from raise pistol and in drawing the pistol from the holster in an emergency.

(5) The instructor shows how to tie the cord to the thumbpiece of the hammer and has each man adjust his cord in the same manner.

(6) Next, the instructor demonstrates cocking of the pistol by means of the cord. He explains how this expedient simulates the self-loading action of the pistol.

(7) The instructor shows how the pistol is kept as nearly on the mark as possible during the whole score. He cautions the men to avoid unnecessary flourishes or movements between shots.

(8) Finally, the instructor explains and demonstrates—

(*a*) The action of the pistol in recoil.

(*b*) Why the arm should not be permitted to bend at the elbow.

(*c*) Moving the pistol upward through a small arc and deflecting it only a short distance from the original point of aim.

(*d*) Moving the forefinger forward after the explosion only far enough to allow the sear to become reengaged and, immediately thereafter, beginning the trigger-squeeze for the next shot.

(*e*) Keeping the eye from closing when the explosion occurs.

(*f*) Holding the breath for each shot.

(9) (*a*) The above demonstrations having been completed, the men are placed in pairs in front of the line of targets, one to practice and one to coach. The exercise is then conducted to simulate sustained fire in range practice. If a line of disappearing tar-

gets has been arranged for this exercise, the targets appear, remain in sight the allotted time, and then disappear. While the targets are in sight, each man undergoing instruction attempts to fire five shots (simulated fire), cocking the piece for each shot, except the first, by a jerk of the cord with the left hand.

(*b*) If the targets are stationary, the exercise begins with the command: 1..COMMENCE, 2. FIRING, and ends with the command: 1. CEASE, 2. FIRING.

(*c*) After each three or four scores of simulated fire, the coaches and pupils exchange duties.

(10) (*a*) In this exercise the coach notes and corrects all errors in grip, position, trigger squeeze, cocking and manipulation of the piece, and holding the breath. He gives particular attention to the trigger squeeze.

(*b*) Sustained-fire exercises should be frequent but not of long duration.

(*c*) It is advisable to extend the time limits several seconds when sustained-fire exercise is first taken up. The time limit is then gradually reduced until it corresponds to the time prescribed for range firing, record practice.

83. SUSTAINED FIRE; REVOLVER. a. Training for sustained fire. Training for sustained fire is begun after the principles of slow fire and timed fire are thoroughly understood and some facility in applying them has been gained.

b. General principles. (1) Sustained fire is the same as slow fire and timed fire except that the piece remains pointed at the target for five consecutive shots, and there is no pause or delay between the discharge of one shot and the application of the operations to fire the next shot.

(2) To attain the degree of accuracy required for proficiency in sustained fire, the revolver must be cocked by use of the thumb and not by using double action.

(3) Basically, the grip on the stock (except for a slightly tighter feel and the fact that the finger tips now contact the stock), the position of the arm, and the position of the body are the same as for slow fire and timed fire.

(4) It is important that a uniform grip be maintained throughout the firing of the score. Any shift in the position of the revolver in the hand or variation in the pressure of the hand and fingers during the firing of the score results in inaccuracy. A great deal of practice "dry shooting" is necessary to develop skill.

(5) To fire the first shot, the revolver is brought from the position of raise pistol by the shortest route to the firing position with the sights properly aligned on the aiming point. This is done by a smooth but deliberate and rapid extension of the right arm straight from the shoulder, cocking the piece, inserting the forefinger in the trigger guard during the movement, and holding the breath.

(6) In sustained fire, time is gained by—

(*a*) Taking position accurately.

(*b*) Applying a heavy initial pressure on the trigger as soon as the sights are aligned and then maintaining a continuously increasing pressure until the shot is fired.

(c) Rapid and smooth cocking of the piece.

(d) Keeping the focus of the eye on the front sight during the firing of the entire string.

(e) Absorbing the shock of recoil at the shoulder, not at the wrist or elbow, thereby reducing the movement of the gun to a minimum.

Caution: *Every effort should be made to overcome the disastrous tendency to save time by pulling the trigger quickly when the aim is perfect.*

c. Cocking revolver. (1) *General.* There are two methods of cocking the revolver for the succeeding shots, the side method and the straight-back method. Both methods are good. Each requires a great deal of practice before sufficient skill is acquired to cock the hammer without shifting the position of the stock in the hand. The method to be used by the individual depends upon the size, shape, and muscular development of the hand. For that reason both methods should be taught and practiced until it is determined which one proves the more satisfactory. Thereafter only that method should be practiced.

(2) *Side method* (fig. 28). (a) The recoil of the revolver causes it to rise about 4 to 6 inches above the point of aim. As it reaches the top of the upward movement, the grip is relaxed slightly and the ball of the thumb is placed on the spur of the hammer. A downward pressure is exerted with the thumb and at the same time the muzzle is moved with a wrist motion to the right about 4 inches. This combined action of the thumb and movement of the muzzle causes the hammer to snap to full cock.

(*b*) During this movement it is important that the fingers on the left side of the stock be kept in place to assist in controlling the revolver.

(*c*) Immediately after the hammer has snapped to the full cock position, the revolver is moved back into aiming position and the thumb replaced along the side of the frame.

(*d*) During the operation of cocking and bringing

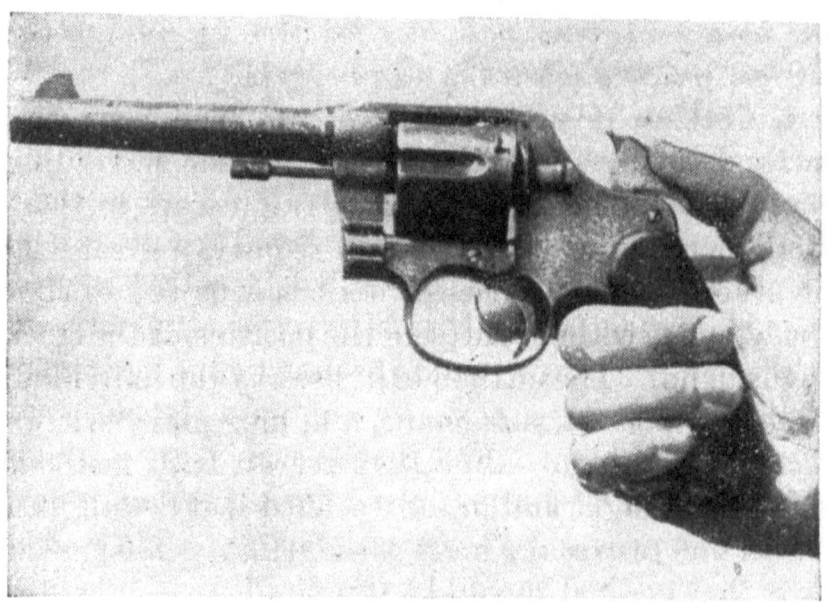

Figure 28. Side method of cocking revolver.

the revolver back to the aiming position, the muzzle must be kept elevated so that the front sight is visible and can be readily aligned in the rear sight notch. This obviates the possibility of losing time in hunting for the front sight and in aligning it in the rear sight notch, which often occurs when the muzzle is allowed to sag.

(*e*) Some individuals may experience difficulty in

keeping the grip in the same position on the revolver during the firing of a string. In most cases the hand tends to work higher on the stock, thus restricting the action of the thumb in cocking and making it necessary to regrasp the revolver in the middle of the string. This difficulty may be overcome by altering the grip slightly so that the little finger is placed under the bottom of the butt of the stock (fig. 29).

(3) *Straight-back method* (fig. 30). (*a*) With this method of cocking the revolver, the grip is not loosened nor is the revolver shifted from its line of recoil. As soon as the shot is fired and while the gun is in recoil, the thumb is placed on the hammer spur and the hammer drawn straight back to the full cock position by the action of the thumb only. During the time the hammer is being drawn back, the revolver is lowered from its uppermost recoil position to the aiming position. As soon as the hammer is cocked, the thumb is replaced alongside the frame.

(*b*) This method is simpler and has the following advantages over the side method:

> *1.* Permits the grip to be more uniformly maintained throughout the firing of the sustained-fire string.
> *2.* Since there is no side movement of the revolver during the process of cocking, the sights can be more readily realigned and brought back on the point of aim.

(*c*) The straight back method has the disadvantage, however, that many men are unable to flex their thumbs sufficiently to draw the hammer all the way back. This causes the thumb to become cramped

Figure 29. Cocking the revolver, straight-back method, little finger beneath butt of stock.

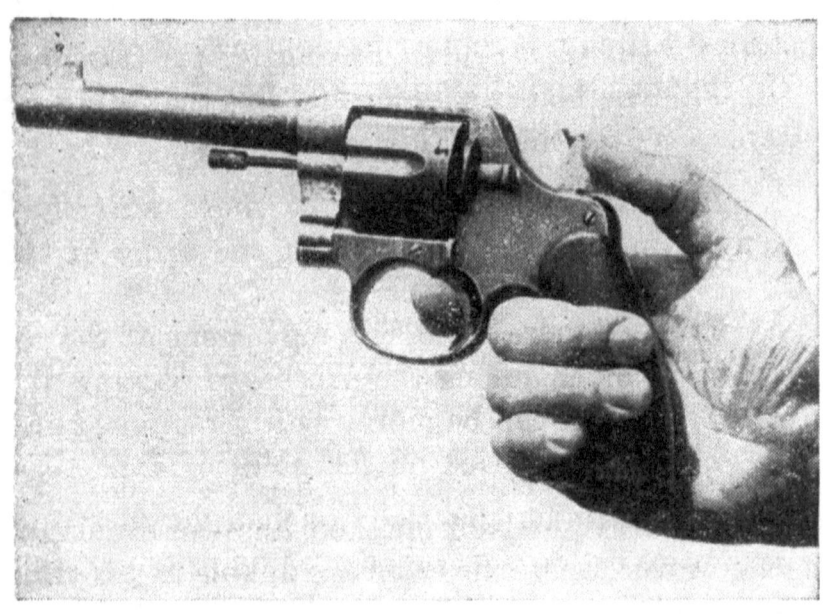

Figure 30. Cocking the revolver, straight-back method.

when the hammer is about two-thirds of the way to full cock, and necessitates regripping the piece to complete the cocking of the hammer. This is a very bad feature as it results in a loss of time and cadence in firing the string. This difficulty may, in some cases, be obviated by slightly lowering the grip, although care should be exercised that the grip is not too low, as this results in the stock sliding further into the hand with each shot fired. Men who experience this difficulty should be required to use the side method of cocking the hammer.

d. Cocking exercise. (1) This exercise is held for the purpose of acquiring a smooth and rapid cocking operation.

(2) Practice in cocking the revolver should be conducted employing both of the above-described methods in order to enable the man to determine which method is best adapted to the conformation and flexibility of his hand. Thereafter, only the method selected is practiced. Before being considered proficient, the man must be able to cock the revolver at least 15 times in 10 seconds. The first hour of rapid-fire training should be devoted to cocking exercises. Thereafter, each pupil should be given additional practice from time to time until he is considered proficient.

(3) The exercise is conducted by the coach-and-pupil method. The instructor explains and demonstrates both methods. Emphasis is placed on the following points: The importance of maintaining a uniform grip; the necessity of keeping the muzzle of the revolver high; and the advantages of keeping the eye focused on the front sight. This exercise

should not be continued longer than about 10 seconds at a time. Frequent changes of coach and pupil are necessary to prevent undue tiring of the muscles of the arm and hand. After requiring the pupil and his coach to take position on the line, the instructor commands: 1. COCKING EXERCISE—READY, 2. EXERCISE, 3. CEASE FIRING, 4. REST. At the first command the pupil grasps the grip as described in paragraph 79 and extends the revolver to the firing position. At the second command the eye is focused on the front sight, without attempting to aim, and the trigger is squeezed causing the hammer to fall. The operations of cocking the hammer by one of the prescribed methods are then continued until CEASE FIRING is given, when the revolver is brought to the position of RAISE PISTOL. At the command REST, the revolver is returned to RAISE PISTOL while the hand and arm are rested.

e. Sustained-fire exercise. (1) Required for this exercise: A row of L targets or a row of aiming bull's-eyes.

(2) Give the command: 1. INSPECTION, 2. ARMS, and verify the fact that all revolvers are unloaded.

(3) Explain to the assembled command that the trigger squeeze is the same in sustained fire as in slow fire.

(4) Demonstrate the correct method of bringing the revolver by the shortest route to the aiming position. Show how this is done from RAISE PISTOL and in drawing the revolver from the holster in an emergency.

(5) Show how the revolver is held to facilitate

rapid cocking with the right thumb without disarranging the hold.

(6) Demonstrate the method of cocking the revolver with the right thumb, without bending the elbow.

(7) Show how the revolver is kept as nearly on the mark as possible during the whole score. Caution the men to avoid unnecessary flourishes or movements between shots.

(8) Demonstrate—

(*a*) The action of the revolver in recoil when a shot is fired.

(*b*) How the arm should not be permitted to bend at the elbow.

(*c*) How the revolver should move upward through a small arc and be deflected from the original point of aim only a short distance.

(*d*) How the forefinger should move forward after the explosion only far enough to allow the trigger to become reengaged with the hammer and immediately after cocking start pressing the trigger for the next shot.

(*e*) How the eye should not be allowed to close when the explosion occurs.

(*f*) How the breath should be held for each shot.

(9) (*a*) The above demonstrations having been completed, the men are placed in front of the line of targets in pairs, one to practice and one to coach. The exercise is then carried on exactly the same as sustained fire in range practice. If a line of disappearing targets has been arranged for this exercise, the targets appear, remain in sight the allotted time, and then disappear. While the targets are in sight,

each man undergoing instruction attempts to fire five shots (simulated fire), cocking the piece for each shot except the first with the thumb.

(*b*) If the targets are stationary the exercise begins with the command: 1. COMMENCE, 2. FIRING, and ends with the command: 1. CEASE, 2. FIRING.

(*c*) After each three or four scores of simulated fire the men of each pair are directed to change places, the firer becoming the coach and the coach becoming the firer.

(10) (*a*) **In** this exercise the coach carefully watches the man and corrects all errors in grip, position, trigger squeeze, cocking and manipulation of the piece, paying particular attention to the trigger squeeze.

(*b*) Sustained-fire exercises should be frequent but not of long duration.

(*c*) It is advisable to extend the time limits several seconds when sustained-fire exercise is first taken up. The time limit is then gradually reduced until it corresponds to the time prescribed for range firing, record practice.

84. QUICK FIRE. a. Training for quick fire. (1) Training in quick fire is taken up after the sustained-fire exercise has been practiced sufficiently to be understood thoroughly. Exercises in slow fire, timed fire, sustained fire, and quick fire should be continued until the end of the period of preparatory training.

(2) For each shot, the pistol is brought from RAISED PISTOL to the aiming position by the shortest route after the target appears.

(3) The automatic pistol may be cocked between

shots by means of a cord (as in sustained-fire exercise), or by using the left hand to pull the hammer back after the position of raise pistol is resumed. The revolver is cocked with the thumb *after* each shot but *before* the position of raise pistol is resumed.

b. Quick-fire exercise. (1) Required for this exercise: A line of E targets that can be operated as bobbing targets from a pit or screen, or a line of E targets so arranged on pivots that the edge can be turned toward the firer when the target is not exposed.

(2) Give the command: 1. INSPECTION, 2. ARMS, and verify the fact that all revolvers are unloaded.

(3) Explain to the assembled command that the trigger squeeze is the same in quick fire as in slow fire.

(4) Demonstrate the correct method of bringing the piece from raise pistol to the aiming position.

(5) Show how the pistol is cocked between shots.

(6) The above demonstrations having been completed, the men are placed in pairs in front of the line of bobbing targets, one man of each pair to act as coach for the other man. The exercise is then carried on exactly the same as quick fire in range practice. The targets appear, remain in sight the allotted time, and then disappear. After the targets appear each man undergoing instruction brings his pistol from raise pistol to the aiming position, aims, fires one shot (simulated fire), and returns his piece to the position of raise pistol. After three or four scores of simulated fire the men of each pair are directed to change places.

(7) The coach watches carefully the man going

through the exercises and corrects all errors in the grip, position, holding the breath, trigger squeeze, and the manipulation of the piece, paying particular attention to the trigger squeeze. It is advisable to extend the time limit about two seconds for each shot when quick-firing exercise is first taken up. The time is then gradually reduced until it corresponds to the time prescribed for range firing, record practice.

(8) When disappearing targets cannot be provided for this exercise it may be held with stationary E targets. The command UP is given to signify that the targets are in sight, and the command DOWN to signify that they have been withdrawn.

(9) Practice in quick fire should be held frequently, but the periods of practice should not be of long duration.

(10) If the range is some distance from the area designated for preparatory exercises, or it is impracticable to arrange for a line of bobbing targets, L targets may be substituted for the bobbing targets.

85. EXAMINATION; AUTOMATIC PISTOL. At the completion of the preparatory instruction, the instructor should assure himself by an examination that every man understands thoroughly and can explain every phase of the preparatory training. The questions and answers given below are merely examples. Each man should be required to explain each item in his own words.

Instructor: Examine your pistol to see that it is unloaded.

Q. What are the safety devices of the pistol?

A. The safety lock, the grip safety, the half-cock notch, and the disconnector.

Q. Show me how you test the safety lock.

A. I cock the pistol, move the safety lock up into place, and then grip the stock and see if the hammer remains up when pressure is applied to the trigger.

Q. Show me how you test the grip safety.

A. I cock the pistol, see that the safety lock is down and then, without putting any pressure on the grip safety, I see if the hammer will remain up when a strong pressure is applied to the trigger.

Q. Show me how you test the half-cock safety device.

A. I half cock the pistol, grip the stock, and see if the hammer remains at half cock when pressure is applied to the trigger. Then I take my finger off the trigger, pull the hammer back almost to full cock, and let go of it to see if it stops at half cock as it falls.

Q. Show me how you test the disconnector.

A. I cock the pistol and grip the stock; then with my left hand I move the slide to the rear ¼ inch; I then apply a strong pressure on the trigger and release the slide to see if the hammer will remain up. I also pull the slide fully back until it is held in place by the slide stop; I then grip the stock, apply a strong pressure on the trigger and release the slide by pressing down the slide stop with my left hand. The hammer should remain up after the slide moves forward into place.

Q. If the hammer does not remain up after the slide moves forward into place, what does it indicate?

A. That with ball ammunition the pistol would

continue to fire automatically as long as pressure is maintained on the trigger, which is very dangerous.

Q. If any of the tests of the safety devices fail at any time, what should you do?

A. I should report the matter at once to my platoon or company commander.

Q. What is this (indicating a sighting bar)?

A. A sighting bar.

Q. What is it used for?

A. To teach men how to aim.

Q. Why is it better than a pistol for this purpose?

A. Because the sights are much larger and slight errors can be seen more easily and pointed out.

Q. What does this represent?

A. The front sight.

Q. What does this represent?

A. The rear sight.

Q. What is this?

A. The eyepiece.

Q. What is it for?

A. To make the man hold his head in the right place so that he will see the sights properly aligned.

Q. Is there an eyepiece on the pistol?

A. No. A man learns by the sighting bar how the sights look when properly aligned, and he must hold the pistol while aiming so as to see the sights in the same way.

Q. Adjust the sights of this sighting bar so that they are in proper alignment with each other. (Verified by instructor.)

Q. Now that the sights are properly adjusted, have the small bull's-eye moved until the sights are aimed at it properly. (Verified by instructor.)

Q. Tell me what is wrong with this aim. (The instructor now adjusts the sights of the sighting bar on the bull's-eye with various very slight errors, requiring the man to point out the error.)

Q. Show me how you grip the stock of the pistol.

Q. Show me the position you take when you are going to shoot.

Q. How do you squeeze the trigger?

A. I squeeze it with such a steady increase of pressure as not to know exactly when the hammer will fall.

Q. If the sights get slightly out of alignment while you are squeezing the trigger, what do you do?

A. I hold the pressure I have on the trigger and only go on with the increase of pressure when the sights become aligned again.

Q. If you do this, can your shot be a bad one?

A. No.

Q. Why?

A. Because I cannot flinch, for I do not know when to flinch, and the sights will always be lined up with the bull's-eye when the shot is fired because I never increase the pressure on the trigger except when the sights are properly aligned.

Q. When you are practicing in slow fire and your arm becomes unsteady and your aim uncertain, what should you do?

A. I should come back to raise pistol without firing the shot and then try again after a short rest.

Q. If it is impossible for you to hold the pistol very steady, can you still do good shooting?

A. Yes; if I press the trigger properly.

Q. Tell me why that is.

A. Because the natural unsteadiness of the arm

moves the whole pistol and the barrel remains nearly parallel to the line of sight. But if I give the trigger a sudden pressure the front end of the barrel will be thrown out of line with the target, and the bullets will strike far out from the mark.

Q. What causes this deflection of one end of the pistol when the trigger is given a sudden pressure?

A. The sudden pressure itself causes some of it, but most of it is caused by the flinch that always accompanies this kind of a trigger pressure.

Q. What does a man do when he flinches in shooting a pistol?

A. He usually thrusts his hand forward as if trying to meet the shock by suddenly stiffening all his muscles.

Q. Must the trigger always be squeezed slowly in order to do it correctly?

A. No. I squeeze it the same way in sustained fire and quick fire. The time is shorter but the process is the same.

Q. What is meant by calling the shot?

A. To say where you think the bullet will hit as soon as you shoot and before the shot is marked.

Q. How can you do this?

A. By noticing exactly where the sights point at the time the pistol is fired.

Q. If a man cannot call his shot correctly, what does it indcate?

A. That he did not squeeze the trigger properly and consequently did not know where the sights were pointed at the instant the discharge took place.

Q. Show me how you hold your breath while aiming.

Q. Take your pistol. Aim at that bull's-eye and squeeze the trigger a few times, calling the shot each time. (The instructor particularly notes the holding of the breath.)

Q. Show me how you come to a position of aim from RAISE PISTOL.

Q. Show me how you come to the aiming position in drawing the pistol from the holster in an emergency.

Q. Take this pistol with the cord tied to the hammer and fire a sustained-fire score at that target (simulated fire).

Q. Fire a score (simulated fire) at that quick-fire target. I will give the command UP when it is supposed to come into sight, and the command DOWN when it is supposed to be withdrawn from view.

Q. What do you do in case a cartridge misses fire?

A. I bring the piece to RAISE PISTOL, grasp the slide with my left thumb and finger, pull the slide fully back and let go of it. This throws out the faulty cartridge and loads in another cartridge.

Q. Are there any points about pistol firing that you do not understand?

Note. In all the demonstrations by the man undergoing examination the instructor carefully notes all points that are covered in the preparatory exercises. Each man is put through a thorough test along the line indicated in these questions and answers before he is allowed to fire.

86. EXAMINATION; REVOLVER. The examination on the revolver is conducted in the same manner as the examination on the automatic pistol.

Instructor: Examine your revolver to see that it is unloaded.

Q. What are the safety devices on the revolver?
A. The safety and the cylinder bolt.
Q. Show me how you test the safety.
A. I cock the revolver, hold the hammer back with the thumb while pressing the trigger to disengage it from the hammer, let the hammer down a little way, release the trigger, then release the hammer. I see if the hammer falls all the way forward.
Q. Show me how you test the cylinder bolt.
A. With the hammer down I attempt to rotate the cylinder. If it moves more than about $\frac{1}{64}$ inch in either direction the revolver is faulty.
Q. If the tests of the safety devices fail at any time, what should you do?
A. I should report the matter at once to my platoon or company commander.
Q. Show me how to load the Colt revolver with three (five) loose rounds so that it will fire the first time it is cocked and each time thereafter.
A. I open the cylinder and insert three (five) cartridges in consecutive chambers. I then close the cylinder, pull back the hammer to about one-fourth full cock position and rotate the cylinder until the first loaded chamber is next on the *left* of the empty chamber aligned with the barrel.
Q. Show me how you let the hammer down on a loaded revolver without firing.
A. I pull the hammer a little to the rear of full cock with the thumb and holding it back I press the trigger, let the hammer forward about $\frac{1}{4}$ inch with the thumb, release the trigger, and then lower the hammer all the way with the thumb.
Q. What is this (indicating a sighting bar)?

A. A sighting bar.

Q. What is it used for?

A. To teach men how to aim.

Q. Why is it better than a revolver for this purpose?

A. Because the sights are much larger and slight errors can be seen more easily and pointed out.

Q. What does this represent?

A. The front sight.

Q. What does this represent?

A. The rear sight.

Q. What is this?

A. The eyepiece.

Q. What is it for?

A. To make the man hold his head in the right place so that he will see the sights properly aligned.

Q. Is there an eyepiece on the revolver?

A. No. A man learns by the sighting bar how the sights look when properly aligned, and he must hold the revolver while aiming so as to see the sights in the same way.

Q. Adjust the sights of this sighting bar so that they are in proper alignment with each other. (Verified by instructor.)

Q. Now that the sights are properly adjusted, have the small bull's-eye moved until the sights are aimed at it properly. (Verified by instructor.)

Q. Tell me what is wrong with this aim. (The instructor now adjusts the sights of the sighting bar on the bull's-eye with various very slight errors, requiring the man to point out the error.)

Q. Show me how you grip the stock of the revolver.

Q. Show me the position you take when you are going to shoot.

Q. How do you squeeze the trigger?

A. I squeeze it with such a steady increase of pressure as not to know exactly when the hammer will fall.

Q. If the sights get slightly out of alignment while you are squeezing the trigger, what do you do?

A. I hold the pressure I have on the trigger and go on with the increase of pressure only when the sights become aligned again.

Q. If you do this can your shot be a bad one?

A. No.

Q. Why?

A. Because I cannot flinch, for I do not know when to flinch, and the sights will always be lined up with the bull's-eye when the shot is fired because I never increase the pressure on the trigger except when the sights are properly aligned.

Q. When you are practicing in slow fire and your arm becomes unsteady and your aim uncertain, what should you do?

A. I should come back to RAISE PISTOL without firing the shot and then try again after a short rest.

Q. If it is impossible for you to hold the revolver steady, can you still do good shooting?

A. Yes; if I press the trigger properly.

Q. Tell me why that is.

A. Because the natural unsteadiness of the arm moves the whole revolver and the barrel remains nearly parallel to the line of sight. But if I give the trigger a sudden pressure the front end of the barrel will be thrown out of line with the target, and the bullets will strike far out from the mark.

Q. What causes this deflection of one end of the revolver when the trigger is given a sudden pressure?

A. The sudden pressure itself causes some of it, but most of it is caused by the flinch that always accompanies this kind of a trigger pressure.

Q. What does a man do when he flinches in shooting a revolver?

A. He usually thrusts his hand forward as if trying to meet the shock by suddenly stiffening all his muscles.

Q. Must the trigger always be squeezed slowly in order to do it correctly?

A. No. I squeeze it the same way in sustained fire and quick fire. The time is shorter but the process is the same.

Q. What is meant by calling the shot?

A. To say where you think the bullet will hit as soon as you shoot and before the shot is marked.

Q. How can you do this?

A. By noticing exactly where the sights point at the time the revolver is fired.

Q. If a man cannot call his shot correctly, what does it indicate?

A. That he did not squeeze the trigger properly and consequently did not know where the sights were pointed at the instant the discharge took place.

Q. Show me how you hold your breath while aiming.

Q. Take your revolver. Aim at that bull's-eye and squeeze the trigger a few times, calling the shot each time. (The instructor particularly notes the holding of the breath.)

Q. Show me how you come to a position of aim from raise pistol.

Q. Show me how you come to the aiming position in drawing the revolver from the holster in an emergency.

Q. Take this revolver and fire a sustained fire score at that target (simulated fire). I will command COMMENCE FIRING to start the score and CEASE FIRING to stop it.

Q. Fire a score (simulated fire) at that quick-fire target. I will give the command UP when it is supposed to come into sight, and the command DOWN when it is supposed to be withdrawn from view.

Q. What do you do in case a cartridge misses fire in combat?

A. I recock the pistol for the next shot.

Q. Are there any points about firing the revolver that you do not understand?

Section III

COURSES TO BE FIRED

87. GENERAL. AR 775-10 prescribes details as to ammunition allowances and personnel who will fire.

88. INSTRUCTION PRACTICE. The following tables prescribe the instruction practice to be fired from the standing position and the order to be followed by the individual firer. Target L, used in much of the practice, stimulates competition and reveals all errors with greater clarity than other types of target.

a. Slow fire.

Table I. Slow fire—Target L.

Range (yards)	Time	Shots
15	No time limit	5
25	do	5

Unlimited time is permitted for slow fire in order to permit proper explanation of the causes of errors and indication of corresponding remedies. It is intended to be the elementary phase of instruction in the proper manipulation of the weapon and for determining and correcting the personal errors of the firer.

b. Time fire.

Table II. Timed fire—Target L

Range (yards)	Time	Shots
15	2 minutes and 30 seconds	5
25	do	5

c. Sustained fire.

Table III. Sustained fire—Target L

| Range (yards) | Time (seconds) | | Shots |
	Automatic pistol	Revolver	
15	12	15	5
25	15	18	5

d. Quick fire.

Table IV. Quick fire—Target E—Bobbing

Range (yards)	Time	Shots
15	3 seconds per shot	5
25	do	5

89. RECORD PRACTICE. The following tables prescribe the firing in record practice to be fired from the standing position and the order to be followed by the individual firer. The procedure is as in instruction practice.

a. Timed fire.

Table V. Timed fire—Target L

Range (yards)	Time	Shots
*25	2 minutes and 30 seconds	5

b. Sustained fire.

Table VI. Sustained fire—Target L

| Range (yards) | Time (seconds) | | Shots |
	Automatic pistol	Revolver	
*15	12	15	5
*25	15	20	5

c. **Quick fire.**

Table VII. Quick fire—Target E—Bobbing

Range (yards)	Time	Shots
*25	3 seconds per shot	5

*Fire twice for one score.

90. SCORING AND CLASSIFICATION. a. The firer's individual score is equal to the total number of points made on all targets. A hit on the E target (bobbing) is evaluated as 5 points and the value of a hit on the L target is determined by its location. The maximum possible scores are: instruction practice, 450; record practice, 350.

b. Individual classification and minimum scores required for qualification in dismounted pistol and revolver marksmanship for each individual authorized or required to complete record practice are as follows:

	Points	
	Pistol	Revolver
Pistol expert	280	285
Pistol sharpshooter	245	250
Pistol marksman	210	215

Section IV

RANGE PRACTICE

91. GENERAL. a. Phases. Range practice is initiated immediately after completion of the preparatory marksmanship training. Range practice is divided into two parts—instruction practice and record practice.

b. Sequence of practice. Instruction practice will be completely prior to the commencement of record practice. When record practice is once begun by an individual, it must be completed before he is permitted to undertake any other practice. However, when the time allotted to range practice is limited, the officer in charge of firing may authorize record firing on the same day as instruction practice.

c. Range personnel. (1) *Officer in charge of firing.* An officer in charge of firing will be designated by the responsible commander. It is desirable that he be the senior officer of the largest organization occupying the range. The officer in charge of firing or his assistant will be present during all firing and will be in charge of the practice and safety precautions on the range.

(2) *Range officer.* The range officer is appointed by the appropriate commander and is responsible to the latter for maintaining and assigning ranges, designating danger zones, and closing roads leading into danger zones. The range officer makes timely arrangements for material and labor to place the ranges in proper condition for range practice. He supervises all necessary repairs. When necessary he pro-

vides range guards and instructs them in the methods to be used for the protection of life and property within the danger area.

(3) *Range noncommissioned officer.* A noncommissioned officer may be detailed by the appropriate commander to assist the range officer in the maintenance of the range.

(4) *Target details.* Commanders of organizations firing will provide such detail of officers, noncommissioned officers, and privates as may be necessary to supervise, operate, and mark the targets used by their organizations.

92. INSTRUCTION PRACTICE. a. Range practice. (1) The object of range practice is to teach the individual to apply with a loaded pistol or revolver the principles of good shooting that have been learned during preparatory exercises.

(2) Each individual, while firing, must have a coach to correct his errors.

(3) Slow-fire practice should be conducted until the person receiving instruction thoroughly understands the principles of good shooting.

(4) Timed-fire practice is a step in pistol and revolver marksmanship training that leads from slow fire, without any time limit, to sustained fire, with a short time limit. It requires the firer to aim and fire in a moderate amount of time.

(5) When sustained fire and quick fire are first taken up, the time limit should be extended a few seconds. The time should then be reduced gradually until the scores are being fired in the time prescribed for record practice.

b. Dummy cartridges. (1) Dummy cartridges are of great value in teaching both slow and rapid fire.

(2) *Dummy cartridges must not be used except on the firing line of the pistol range.* The same precautions are observed as in using service ammunition.

c. Slow fire. (1) The coach stands on the left side of the firer in such a position as to be able to observe the latter's trigger finger, his grip, his eye, and his position. It is the duty of the coach to correct all errors. The coach fills the magazines for the firer and hands them to him. At the beginning of range practice, the magazines should be filled partly with service ammunition and partly with dummy cartridges. The firer must not know how many dummy cartridges are in the magazine or the order in which they are packed. This procedure is also applicable to practice with the revolver.

(2) The object of placing dummy cartridges in the magazine is to show the coach whether or not the individual under instruction is squeezing the trigger correctly and, in case of an improper trigger squeeze, to bring the fact forcibly to the attention of the firer himself. When a loaded cartridge is fired, the flinch is often masked by the recoil of the pistol and the firer is not conscious of having flinched. When the hammer falls on a dummy cartridge which the firer thinks is a ball cartridge, the sudden stiffening of the muscles and the thrusting forward of the hand to meet the shock that does not come are apparent to everybody in the vicinity, including the firer himself. Mixing dummy cartridges with service ammunition causes the firer to

make a determined effort to press the trigger properly for all shots.

(3) The firing of scores with dummy cartridges and service ammunition should not be confined to the early stages of training. It is advisable to conduct practice of this kind each day during the entire period of instruction practice. Many expert pistol shots use this form of practice in training for competitions.

(4) Upon beginning range practice, each man receives instruction as indicated in following subparagraphs. Once he has received this instruction, it is usually not necessary to repeat it during subsequent periods of range practice.

(*a*) Explain the method of grasping the piece.

(*b*) Show the amount of force used in gripping the stock by grasping the pupil's hand, saying: "This is too tight a grip" (gripping his hand very tightly); "This is too loose a grip" (gripping his hand loosely); and "This is the right amount of force to use in gripping the stock" (gripping his hand with the firm but comfortable grip that should be used in shooting).

(*c*) Explain and demonstrate the position of the body, the feet, and the arm, and have the pupil assume this position.

(*d*) Explain the proper method of aiming.

(*e*) Explain that any man can aim and hold well enough for a good score. Require the pupil to assume the proper position and aim at the target with an empty pistol (without attempting to squeeze the trigger) to see how long he can hold the sights on or near the bull's-eye. Explain to him that aiming at the target with an empty gun demonstrates the point at which his bullets will strike provided he presses the trigger properly.

(*f*) Explain the proper method of squeezing the trigger.

(*g*) Require the pupil to aim at the target with an empty pistol and then press the trigger for him several times as described in **d**, below, directing the pupil to call the shot each time the hammer falls (fig. 31).

(*h*) Require the pupil to aim at the target with a loaded pistol and then press the trigger for him as described in **d**, below, directing him to call the shot each time the piece is fired. Fire a score of five shots in this way.

(*i*) Require the pupil to fire a score of five shots squeezing the trigger himself to see if he can do it properly and make as good a score as that made when the coach squeezed the trigger.

d. Squeezing trigger. One method of showing the man under instruction how to squeeze the trigger properly is to have him hold and aim the pistol while the coach squeezes the trigger. This is done in the following manner:

(1) The coach demonstrates to the pupil the value of correct trigger squeeze by placing his hands in the position shown in figures 31 or 32 and pressing with his left thumb on the end of the pupil's trigger finger. The coach cautions the pupil neither to assist nor resist the pressure, but to devote his whole attention to his aim and hold.

(2) The coach must be careful to apply a slow, steady pressure to the finger of the pupil and, at the

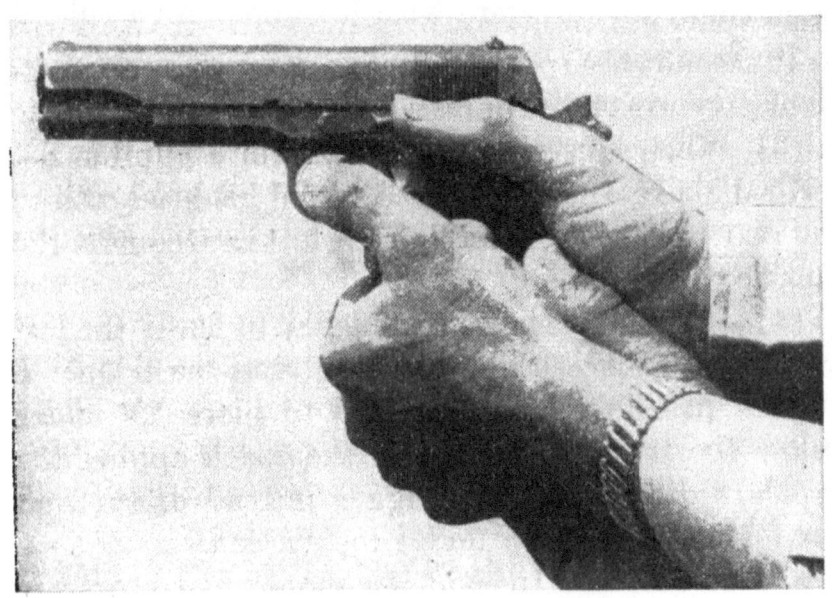

Figure 31. Coach applies pressure to trigger finger of pupil firing the automatic pistol.

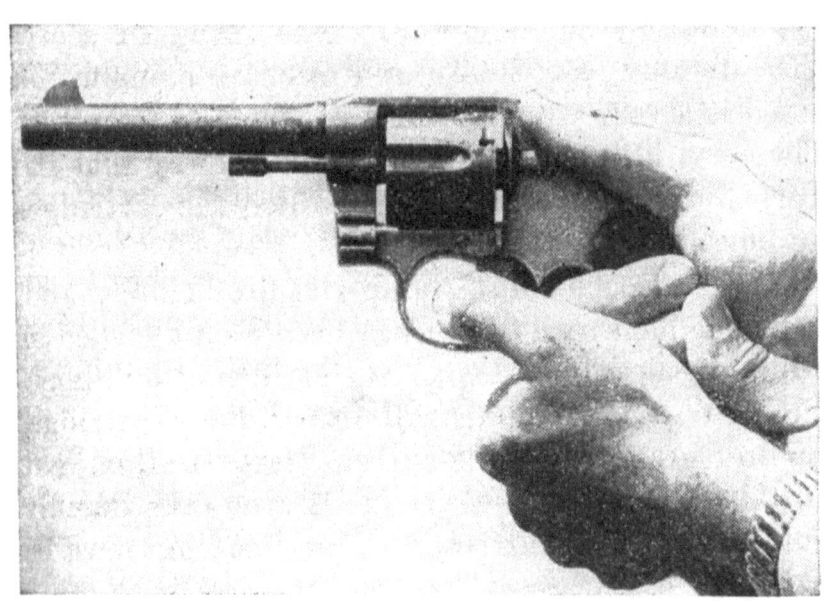

Figure 32. Coach applies pressure to trigger finger of pupil firing the revolver.

same time, not interfere with his aim. As a rule, 5 to 10 seconds are required for the coach to apply sufficient pressure on the pupil's finger to fire the pistol.

(3) When squeezing the trigger for a pupil as described above, the coach should hold his head well to the rear to keep from having his left ear too near the muzzle of the piece.

(4) If the firer shows a tendency to apply the last part of the squeeze himself by giving the trigger a sudden pressure, he is directed to place his finger below the trigger guard, and the coach applies the pressure directly to the trigger instead of through the finger of the man under instruction.

e. Calling shot. Individuals should be required to call each shot in slow and timed fire. If the firer does not call the shot immediately after firing, the coach directs him to do so.

f. Coaching timed fire. (1) The firing of scores with dummy cartridges and service ammunition mixed is a very valuable form of timed-fire practice. The coach fills the magazine in such a way that the firer cannot know the order in which the cartridges are placed.

(2) The coach must watch the firer closely and, each time he is seen to flinch, whether on a loaded or a dummy cartridge, the coach should caution him.

(3) When the hammer falls on a dummy cartridge, the firer grasps the slide with his left hand, pulls it fully back, and releases it. This ejects the dummy and loads another cartridge. The time limit must be extended to compensate for the time lost in ejecting the dummy cartridges. It should not take more than 2 seconds to eject a dummy cartridge and resume

the aiming position. When the revolver is used and the hammer falls on a dummy cartridge, the firer simply cocks his weapon and resumes the aiming position.

g. Coaching sustained fire and quick fire. The use of dummy cartridges and the coaching methods are the same in sustained fire and quick fire as in timed fire.

h. Continued use of dummy cartridges. The occasional use of dummy cartridges in timed fire, sustained fire, and quick fire should be continued throughout the entire period of instruction practice.

93. SAFETY PRECAUTIONS ON THE RANGE. a. Never place a loaded magazine in the automatic pistol nor load the revolver until you have taken your place at the firing point.

b. Always remove the magazine and unload the pistol before leaving the firing point.

c. Always hold the loaded pistol at the position of raise pistol, except while aiming.

d. When firing ceases temporarily, lock the piece and hold it at raise pistol. Do not assume any position except raise pistol without first removing the magazine and unloading.

e. If one or more cartridges remain unfired at the end of a timed-fire, sustained-fire, or quick-fire score, remove the magazine and unload immediately.

f. The range will be kept thoroughly policed at all times. The officer in charge of firing will inspect to insure that empty cartridge cases and unfired rounds are separated from each other and from all trash. Further, he will insure that all empty cartridge cases and unfired rounds are turned in to the range or supply officer immediately upon completion of the day's firing.

94. RANGE ORGANIZATION. a. The work on the range should be so organized that no one is idle for any length of time. A good arrangement is four or six orders per target. It should never be necessary to assign more than six orders. If there is not a sufficient number of targets to allow this, the extra men should remain off the range and be given other instruction.

b. To achieve maximum utilization of facilities and to keep all men fully occupied, the rifle and pistol range may be used simultaneously. To do so, a line of pistol targets is provided about 50 yards to either or both flanks of the rifle range, so arranged that the firing points of both ranges are on one line. The pistol targets may be placed on the ground instead of in pits. The bobbing targets are arranged to revolve on their own axis and are operated from behind the firing-line by means of cords. When the targets are to be marked, the whole line ceases firing, unloads pistols, and moves up to the targets to record the hits and paste patches over the shot holes. In slow fire and timed fire, the coach can keep the firer informed as to the location of his hits by use of field glasses.

c. When the time is short and range facilities and proper supervision permit, one group may fire on the rifle range while the other is firing on the pistol range. As the men complete a score with the rifle, they move to the pistol range and their places at the rifle firing point are filled by men who have completed a score of pistol firing. As soon as all men present have completed their scores with the rifle, the entire group moves back to the next firing point (moving the pistol targets if necessary) and continue as before with the alternate rifle and pistol firing.

d. The pistol targets may be placed so that the line of fire is at right angles to the line of fire of the rifle range if the terrain permits. When it is not practicable to have pistol firing and rifle firing at the same time, other means will be adopted to keep the occupied while they are not actually firing or coaching.

e. Firing at both instruction practice and record practice is conducted in the following manner:

(1) *Slow fire (L target).* (a) *On firing line.*

    1. One firer only will be assigned to a target in each order.

    2. When all firers are in their proper positions on the firing line, the officer in charge commands: LOAD, SLOW FIRE, COMMENCE FIRING. At the latter command the safety is disengaged and the prescribed number of shots are fired at the target. At the conclusion of firing he commands: CEASE FIRING, UNLOAD. No shots will be fired after FIRING.

    3. As the value of each shot is signaled, the scorer announces, in a tone sufficiently loud to be heard by the firer, the firer's name, the number of the shot, and the value of the hit. He then records the value of the hit on the score card.

4. Whenever a target is marked and the firer assigned to that target has not fired, the scorer will notify the officer in charge of firing.

5. When a shot is fired on the wrong target, a miss will not be scored until the target to which the firer is assigned has been withdrawn and the miss has been signaled from the pit.

6. When targets are not operated from a pit, the method of conducting slow fire will be modified as necessary.

(*b*) *In pit.*

1. When a shot is fired on a target, it is withdrawn. A spotter is then placed in the shot hole. The previous shot hole, if any, is pasted, and the target is run up, and the value of the last shot is indicated by a pointer directed at the proper numeral in the vertical margins of the target.

2. When a target frame is used as a counterweight for a double sliding target, the blank side of the frame will be toward the firing line.

(2) *Timed fire (L target).* (*a*) *On firing line.*

1. One firer only will be assigned to a target in each order.

2. When all is ready in the pit, a red flag is displayed at the center target. At that signal the officer in charge of the firing line commands: LOAD. Pistols (revolvers) are loaded.

3. The officer in charge of the firing line then commands: READY ON THE RIGHT, READY ON THE LEFT. Any firer who is not ready calls out, "Not ready on No. —."

4. All being ready on the firing line, the officer in charge commands: READY ON THE FIRING LINE. The safety on each weapon is disengaged and the firers remain at raise pistol. The telephone orderly at the control telephone notifies the pit, "Ready on the firing line."

5. The flag at the center target is waved and then withdrawn. Five seconds after the flag is withdrawn the targets appear, remain fully exposed for the prescribed period of time, and are then withdrawn. Each firer aims at his target as soon as it appears and fires, or attempts to fire, the prescribed number of shots. If a firer fails to fire at all, he will be given another opportunity to fire, but if he fires any shots, the score must stand as his record. He will not be permitted to repeat his score on the claim that he was not ready.

6. As soon as the targets are withdrawn, the officer in charge commands: UNLOAD. Men remain on the firing line until they are ordered off by the officer in charge of firing.

7. As the value of each shot is indicated from

the pits, it is announced by the scorer at the firing line. For example, a score of 5 shots is announced as follows as each shot is marked: Target 10—1 ten, 2 tens; 1 eight, 2 eights; 1 miss. The scorer notes these values on a piece of paper and watches the target as he calls the shot. After marking is finished, he counts the number of shots marked, and if there are more than the prescribed number, he calls "Remark No. —." If the correct number of shots have been marked, he enters the individual value of each hit and the total value of all hits on the firer's score card.

8. When targets are not operated from a pit, the method of conducting timed fire will be modified as follows:

   (a) After the command READY ON THE FIRING LINE has been given by the officer in charge of firing, the safety on each weapon is disengaged and the firers remain at raise pistol. The officer in charge of the firing line then commands: COMMENCE FIRING. At the command FIRING, each firer aims at his target and fires, or attempts to fire, the prescribed number of shots.

   (b) The time is regulated by the officer in charge of the firing line.

   (c) The officer in charge of the firing

line commands: CEASE FIRING, so as to give FIRING at the exact instant that the prescribed time expires. No shots will be fired after FIRING. A short blast of the whistle may be substituted for the latter command.

(d) Other modifications may be made as are necessary.

(b) In pit.
1. The time is regulated by the officer in charge of the pit.
2. When all is ready in the pit, the targets are fully withdrawn and a red flag is displayed at the center target.
3. When the message is received that the firing line is ready, the red flag at the center target is waved and withdrawn, and the command READY is given to the pit details.
4. Five seconds after the red flag is withdrawn, the targets are run up by command or signal, left fully exposed for the prescribed period of time, and then withdrawn.
5. The men in the pit detail examine the targets and put spotters in the shot holes. The targets are then raised and marked.
6. The targets are left up for about 1 minute after being marked, are then withdrawn, pasted, and made ready for another score. They may be left up until ordered pasted by the officer in charge of the firing line.

7. If more than the maximum number of prescribed hits are found on any target, it will not be marked unless all of the hits have the same value. The officer in charge of the firing line will be notified of the fact by telephone.

(3) *Sustained fire* (*L target*). Sustained fire is conducted in the same manner as timed fire, the only difference being in the shorter period of time allowed for the firing of the former.

(4) *Quick fire* (*E target*). (*a*) One firer only will be assigned to a target in each order.

(*b*) When all fires are in their proper positions on the firing line, the officer in charge of the firing line has the target operators turn the edges of the E targets toward the firers. He then commands: LOAD.

(*c*) The officer in charge of the firing line then calls: READY ON THE RIGHT, READY ON THE LEFT. Any firer who is not ready calls out, "Not ready on No. —."

(*d*) All being ready on the firing line, the officer in charge commands: READY ON THE FIRING LINE. The safety on each weapon is disengaged and the firers remain at raise pistol.

(*e*) The officer in charge of the firing line then signals for the targets to be exposed to the firers. As soon as the front of the targets begin to turn toward the firers they aim at the target and fire, or attempt to fire, one shot. Pistols are returned to raise pistol as soon as the one shot is fired or the edge of the targets are again turned toward the firers.

(*f*) The officer in charge of the firing line signals a second time to cause the edge of the targets to be turned toward the firers when the targets have been exposed the correct length of time. After a uniform interval of several seconds he again signals for the targets to be exposed. This procedure is continued until the targets have been exposed the required number of times. Each time the targets are exposed firers aim and fire one shot and return to raise pistol immediately thereafter. The officer in charge of firing then commands: UNLOAD. Men remain on the firing line until they are ordered off by the officer in charge of firing.

(*g*) On orders from the officer in charge of the firing line, firers and scorers move to the targets to count the number of hits on each. The scorers announce the number of hits on each target, in a tone sufficiently loud to be heard by the firer and make the correct entry on the score card.

(*h*) Upon the conclusion of the scoring the officer in charge of firing directs that the targets be pasted and made ready for another score.

(*i*) If more than the maximum number of prescribed hits are found on any target, the firer will be credited with only the number fired, which will not exceed the number prescribed.

95. TARGET DETAILS. a. One commissioned officer is assigned to each two targets. When it is impracticable to assign one officer to each two targets in the pit, an officer will be assigned to supervise the working and scoring of not to exceed four targets. When pit records are prescribed, the officer will take up and sign each score card as soon as a complete score is recorded.

b. One noncommissioned officer is assigned each target to direct and supervise the detail marking and pasting the target. When pit record cards are prescribed, the noncommissioned officer in charge of each target will enter each score on the pit record card and turn it over to the officer in charge of his target. This noncommissioned officer will be selected, except at a one-company post, from an organization other than the one firing on the target which he supervises. When the post is garrisoned by a single company and it is impossible to detail noncommissioned officers of other companies to supervise the marking and scoring, those duties are performed by the noncommissioned officers of the firing company.

c. One or two privates operate, mark, and paste each target.

d. When the targets are not placed in pits, target details may be reduced to one commissioned officer for each four targets, one noncommissioned officer for not to exceed each two targets, and one private to each target.

e. The noncommissioned officer examines the target after each score is fired, enters the score on the score card, and signs thereon his initials. He directs the private to paste the target after the score is recorded and marked and examines the target to see that no shot holes are left unpasted.

96. REGULATIONS GOVERNING RECORD PRACTICE.
a. Coaching. (1) Coaching is permitted during record practice in the period for which no additional

compensation for arms qualification is authorized. The coach may not touch the person or the weapon of the firer. Each firer must observe the location of his own hits.

(2) When additional compensation for arms qualification is authorized, coaching of any nature, after any firer takes his place in the firing point is prohibited. No person may render or attempt to render the firer any assistance while he is taking his position or after he has taken his position at the firing point. Each firer must observe the location of his own hits.

b. Shelter for firer. Sheds or shelter for the firer are not permitted on any range.

c. Cleaning. Cleaning is permitted only between scores.

d. Gloves. A glove may be worn on either or both hands.

e. Weapons loaded on command. Weapons are not loaded except by command or until position for firing has been taken.

f. Shots cutting edge of bull's-eye or lines. Any shot cutting the edge of the figure or bull's-eye is signaled and recorded as a hit in the figure or the bull's-eye. Because the limiting line of each division of the target is the outer edge of the line separating it from the exterior division, a shot touching this line is signaled and recorded as a hit in the higher division.

g. Slow-fire score interrupted. If a slow-fire score is interrupted through no fault of the person firing, the unfired shots necessary to complete the score are fired at the first opportunity thereafter.

h. Misses. In all firing, before a miss is recorded, the target is examined carefully by an officer.

i. Accidental discharges. All shots fired by the firer when it is his turn to fire, the target is ready, and he has taken his place on the firing line, are considered in his score even if his piece is not directed toward the target or is accidentally discharged.

j. Firing on wrong target. Shots fired upon the wrong target are entered as misses upon the score of the individual firing, no matter what the value of the hits upon the wrong target may be. In rapid fire, the individual at fault is credited with only such hits as he may have made on his own target.

k. Two shots on same target. In slow fire, if two shots strike a target at the same time or nearly the same time, and if one of these shots was fired from the firing point assigned to that target, the hit having the highest of the two values is entered on the soldier's score and no record is made of the other hit.

l. Withdrawing target prematurely. In slow fire, if the target is withdrawn from the firing position just as the shot is fired, the scorer at that firing point at once reports the fact to the officer in charge of the scoring on that target. That officer investigates to see if the case is as represented. Being satisfied that such is the case, he directs that the shot be not considered and that the individual fire another shot.

m. Misfires. In case of a misfire in timed fire, sustained fire, or quick fire, the soldier ceases firing and takes the position of raise pistol. In timed fire, the target is not marked and the score is continued after the defective cartridge has been replaced. In sustained fire, the target is not marked and the score is repeated. In quick fire, the score is continued after the defective cartridge has been replaced.

n. Unused cartridges in timed fire and sustained fire. Each unfired cartridge is recorded as a miss.

o. Disabled pistol. If, during the firing of a timed-, sustained-, or quick-fire score, the pistol becomes disabled through no fault of the firer, the procedure outlined in subparagraph **m** above is followed.

p. More than five shots in timed fire or sustained fire. When a target has more than five hits in timed fire or sustained fire, the target is not marked except when all the hits have the same value. In this event, the target is marked and the firer given that value for each shot fired by him.

q. Score cards and scoring. (1) A score card will be kept at the firing point. Entries on all score cards will be made in ink or with indelible pencil. No alteration or correction will be made on the card except by the organization commander or officer scorer in the pit, who will initial each alteration or correction made. The cards at the firing point will bear the date, the firer's name, the number of the target, and the order of firing.

(2) Likewise, when firing is being conducted for compensation for qualification in arms, the scores at each firing point will be kept by a noncommissioned officer of some organization other than that firing on the target to which he is assigned. If this is not possible, company officers will exercise care to insure correct scoring. As soon as a score is completed, the score card will be signed by the scorer, taken up and signed by the officer supervising the scoring, and turned over to the firer's organization commander.

Except when required for entering new scores on the range, score cards will be retained in the personal possession of the organization commander.

(3) In the pit, the officer keeps the scores for the targets to which he is assigned. As soon as a score is completed, he signs the score card. He turns over cards to the organization commander at the end of the day's firing or at such times as requested.

(4) Upon completion of record firing, and after the qualification order is issued, all score cards in his possession will be destroyed by the organization commander.

Section V

KNOWN-DISTANCE TARGETS AND RANGES

97. TARGETS. a. Target E. Target E is a drab silhouette about the height of a soldier in a kneeling position made of bookbinder's board or other similar material (fig. 33). Hits are valued at 1. Any shot cutting the edge of the target is a hit.

b. Target E, bobbing. Target E, bobbing, is so arranged as to be fully exposed to the firer for a limited time; edge of target is toward firer when target is not exposed (fig. 34).

c. Target L. Target L is a rectangle 6 feet high and 4 feet wide, with black circular bull's-eye 5 inches in diameter and seven outer rings (fig. 35). Value of hits in bull's-eye is 10. The diameter of each ring and value of hits are as follows:

Diameter	Value of hit
8½ inches	9
12 inches	8
15½ inches	7
19 inches	6
22½ inches	5
26 inches	4
46 inches	3
Outer, remainder of target	2

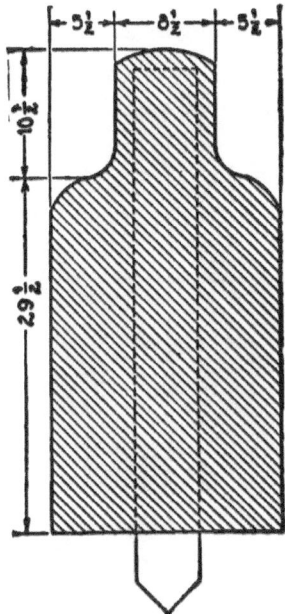

Figure 33. Target E.

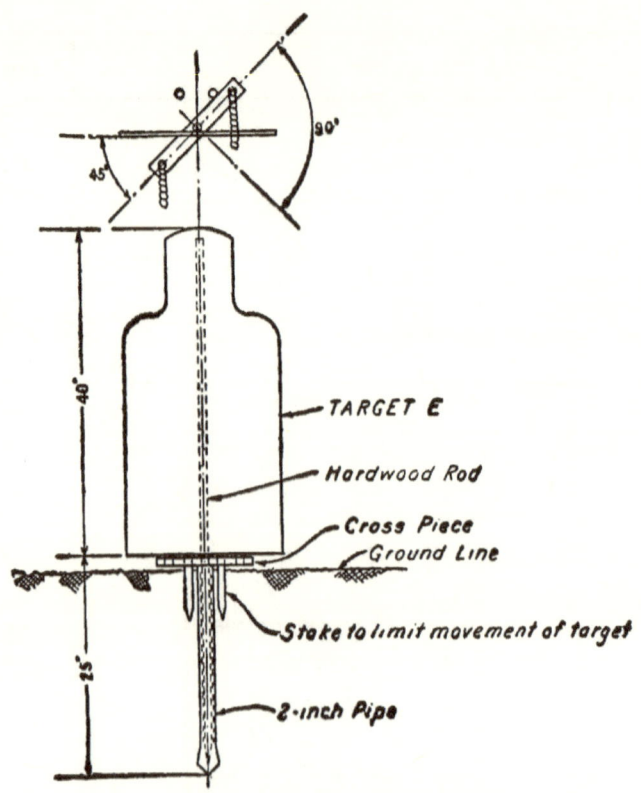

Figure 34. Target E, bobbing.

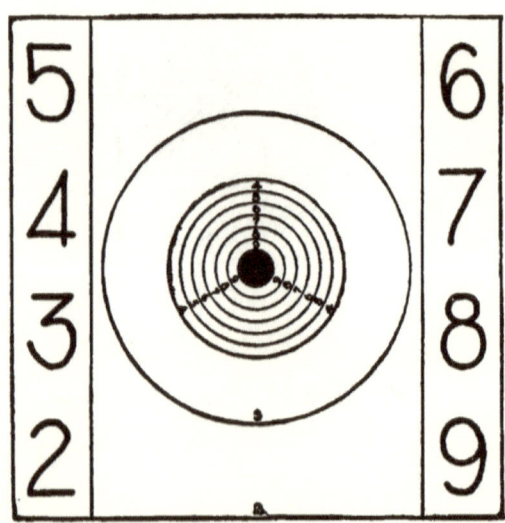

Figure 35. Pistol target L; six feet high and four feet wide.

d. Small-bore targets. No specific targets are prescribed for small-bore practice with the pistol and any suitable targets may be used. The following targets issued by the Ordnance Department are suitable and may be used: Rifle—SB-A-2, SB-A-3, and SB-B-5.

98. RANGES. a. General. Class A target ranges for the rifle, as described in FM 23-5, may be used for pistol practice. The pistol target L may be placed on the sliding target carriage for slow, timed, and sustained fire. Bobbing targets are not ordinarily placed in the pits of rifle ranges, but are set up nearby. If sufficient space is available, an area other than rifle ranges is used for pistol practice.

b. Rules for selection. As the nature and extent of the ground available for pistol practice and also the general climatic conditions are often widely dissimilar at different military posts, it is impossible to prescribe any particular rules governing the selection of ranges. Only certain general conditions may be expressed to which ranges should be made to conform.

c. Safety necessary. For posts situated in thickly settled localities where the extent of military reservation is limited, the first condition to be fulfilled is that of safety for those living or working near, or passing by the range. This requirement can be secured by selecting ground where a natural butt is available or by making an artificial butt sufficiently extensive to stop wild shots. For complete safety, roads, buildings, or cultivated ground should not be nearer than 300 yards to either flank of the range, nor nearer than 1,600 yards to its rear.

d. Direction of the range. If practicable a range should be located so that the direction of firing is toward or slightly to the east of north. Such location gives a good light on the face of the targets during the greater part of the day. However, safety and suitable ground are more important than direction.

e. Best ground for range. Smooth, level ground or ground with only a very moderate slope is best adapted for a range. The target should be on the same level with the firer or only slightly above him, but should not be below.

f. Size of range. The size of the range is determined by its plan and by the number of troops that will fire on it at the same time.

99. PRINCIPLES GOVERNING CONSTRUCTION. a. Intervals between targets. Intervals between targets are usually equal to the width of the targets themselves. When the necessity exists for as many targets as possible in a limited space, this interval may be reduced. Bobbing targets should be placed a minimum of 5 yards apart.

b. Protection for markers. When pits are not used, markers remain in rear of the firing line except during cessation of fire when their duties require them to move to the targets.

c. Artificial butts. If an artificial butt is constructed as a bullet stop, it should be of earth not less than 30 feet high with a slope of not less than 45°. It should extend about 5 yards beyond the outside targets and should be placed as close behind the targets as possible. The slope should be sodded.

d. Hills as butts. A natural hill to form an effective butt should have a slope of not less than 45°. If originally more gradual it should be cut into steps, the face of each step having that slope.

e. Number of targets. Each target should be designated by a number.

f. Measuring range. The range should be carefully measured and marked with a stake in front of each target at each firing point. The stakes should be about 12 inches above the ground and painted white. These stakes then designate the firing points for the different targets at the different distances. Particular care should be taken that each stake thus placed is at right angles to the face of its own target.

g. Danger signals. One or more danger signals are placed near the range to warn passersby when firing is in progress. They should be placed on the roads or on the crest of the hill where they can be plainly seen by those passing.

Section VI

SMALL-BORE PRACTICE

100. GENERAL. Where facilities and equipment permit, all individuals who have satisfactorily passed the examination on preparatory exercises should be advanced to small-bore practice before taking up range practice. There is no recoil or loud report to induce nervousness or flinching, and the soldier soon learns that he can make good scores if he observes the methods and precautions in which he has been instructed. Small-bore practice is not only valuable to the beginner but it affords to the good shot a means of retaining his efficiency.

101. OBJECT. The object of small-bore practice is to provide a form of marksmanship training with the caliber .22 pistol and ammunition which represents the application of the principles taught in the preparatory exercises. Small-bore practice provides an excellent means of improving the standard of shooting in organizations and sustaining interest in marksmanship. Every effort should be made by all organizations to fire the small-bore course prior to regular marksmanship training. The firing of this course enables the organization commander to visualize the state of training of his command and to concentrate his efforts on the training of those who are most deficient.

102. CONTINUOUS PRACTICE. Small-bore practice should be carried on throughout the year subject to such limitations as may be imposed by available ammunition and range facilities. All persons who have never been properly instructed in shooting methods prescribed herein should be given a thorough course of preparatory instruction before being permitted to fire on the small-bore range. All small-bore practice is organized and supervised in accordance with the methods of instruction as prescribed in this manual.

103. SMALL-BORE PRACTICE COURSE. Prior to range firing, an instruction course and a record course, similar to those prescribed in paragraphs 88 and 89, should be fired with the caliber .22 pistol by each individual. No reports of the results of small-bore practice are required but the firing record of individuals should be posted in order to stimulate interest and competition among the men of the organizations.

104. ADDITIONAL PRACTICE. In addition to the practice course indicated in paragraph 103, small-bore practice should be carried on throughout the year. The amount and details of the practice are left to the discretion of the organization commander. Varied targets such as tin cans, bottles, pendulums, and moving targets stimulate interest. Matches between individuals and teams of the same or different units should be promoted.

CHAPTER 2

COMBAT FIRING

Section I

GENERAL

105. GENERAL. a Definition. Combat firing with hand weapons is range practice conducted under conditions which approximate those of combat as nearly as practicable.

b. Objective. The objective of instruction and practice in combat firing is to develop proficiency of the individual in firing the pistol or revolver quickly and accurately under combat conditions. In order to attain this objective, training in the positions and simulated firing exercises should be carried on throughout the year.

c. Scope. Combat firing includes instruction and practice in dismounted positions, simulated firing at different objects at all gaits while mounted, further application of principles learned in basic marksmanship, and range firing.

d. Prior training requirements. (1) Individuals undertaking the course of instruction in combat firing should have first completed all phases of training prescribed herein for preparatory marksmanship, practice firing, and record firing.

(2) Training in marksmanship, including attendant range firing periods, is conducted under conditions which are advantageous to the firer. Distracting influences such as uncertain footing, unsteady targets, or time limitations, are minimized in order

that the soldier may learn the principles of good marksmanship and develop proficiency in their practice. Firing in combat, either mounted or dismounted, is vastly different from firing on a well-constructed range at a fixed target which cannot return the fire. Within the limits of safety and common sense, combat firing should duplicate for the soldier the difficulties he is apt to encounter when firing at an armed enemy in battle. It is obviously impracticable to attempt to reproduce exactly combat conditions on any range, particularly the mental and physical tension and reactions of the individual under fire. However, thorough instruction and properly supervised combat firing will develop the soldier's ability to fire instinctively, accurately, and quickly.

Section II

DISMOUNTED

106. GRASPING THE WEAPON. a. One-hand grip. The manner of grasping the revolver or pistol described in paragraph 79 is also applicable to dismounted combat firing.

b. Two-hand grip. Use of the two-hand grip enables the firer to support the one-hand grip thereby attaining more consistent accuracy. The two-hand grip is as follows: First, the weapon is grasped in the right hand in the usual manner. The butt of the weapon, still grasped in the right hand, is then seated firmly in the palm of the left hand. The fingers and thumb of the left hand are then closed over the right hand in a manner that will provide maximum support (fig. 36). The two-hand grip should duplicate as nearly as possible the support which is provided when an automobile hood, fence post, or other object is utilized as a rest during combat.

Figure 36. The two-hand grip.

107. POSITIONS. a. Prone. The soldier lies as flat as possible on the ground with legs apart and heels down in the same manner as when firing the rifle. The head and body are on a line with the target toward which both arms are fully extended with the pistol grasped in the two-hand grip (fig. 37). The prone position is employed when firing at a range of 50 yards or more.

Figure 37. Prone position; two-hand grip.

b. Kneeling. The kneeling position is most advantageously employed at ranges of 25 to 50 yards. It is similar to the kneeling position used when firing the rifle. The firer kneels on the right knee and rests his left upper arm on the raised left knee with the elbow projecting beyond the knee. The butt of the pistol, grasped in the right hand, is then seated in the palm of the left hand and the two-hand grip assumed (fig. 38). Men who shoot left handed may reverse the position.

Figure 38. Kneeling position; two-hand grip.

c. Standing. In the standing position, the body is in a forward crouch with the knees flexed and the trunk bent forward from the hips (fig. 39). The feet are placed naturally in a position which will allow another step toward the target. At all times the body should be maintained in a comfortably balanced position facilitating rapid movement in any direction. The standing position is used when engaging targets at ranges of about 15 yards, particularly surprise targets.

d. Ready position. In the ready position, the pistol is grasped in the right hand in the usual manner.

The gun arm is flexed with the forearm in a horizontal position. The upper arm is held close to the body with the elbow resting comfortably against the hip bone.

108. COORDINATION OF MOVEMENT. In combat firing it is essential that the individual have instant coordination between the eyes, ears, brain, and muscles. Immediately upon perceiving or suspecting a target, he must assume the best position indicated by the situation and aim and fire, or be prepared to fire. The firer must shoot quickly and instinctively without the deliberate use of sights, except at the longer ranges of 25 to 50 yards.

   a. Prone and kneeling. When in the ready position and a target appears at ranges over 25 yards, the firer assumes the prone or kneeling position as if he were using the rifle. At ranges of 50 yards or more, accurate fire can be delivered by employing the prone position and the two-hand grip to facilitate sighting and aiming. At ranges 25 to 50 yards, the kneeling position provides additional support necessary to fire rapidly and accurately.

   b. Standing. (1) The standing position is the position most frequently used in combat, consequently, the one to which the greatest amount of practice should be given. When standing or walking with the pistol held in the ready position, the firer takes an additional step toward targets appearing at ranges of less than 25 yards. Simultaneously, he thrusts the pistol toward the target as if he were pointing his finger in accusation (fig. 39). The first shot may be fired as the gun arm is being extended. The second shot is fired with the arm fully extended and the wrist and elbow locked, with the only pivot point being the shoulder. Despite the omission of careful aiming, a reasonable amount of practice will develop accuracy in engaging targets at short ranges.

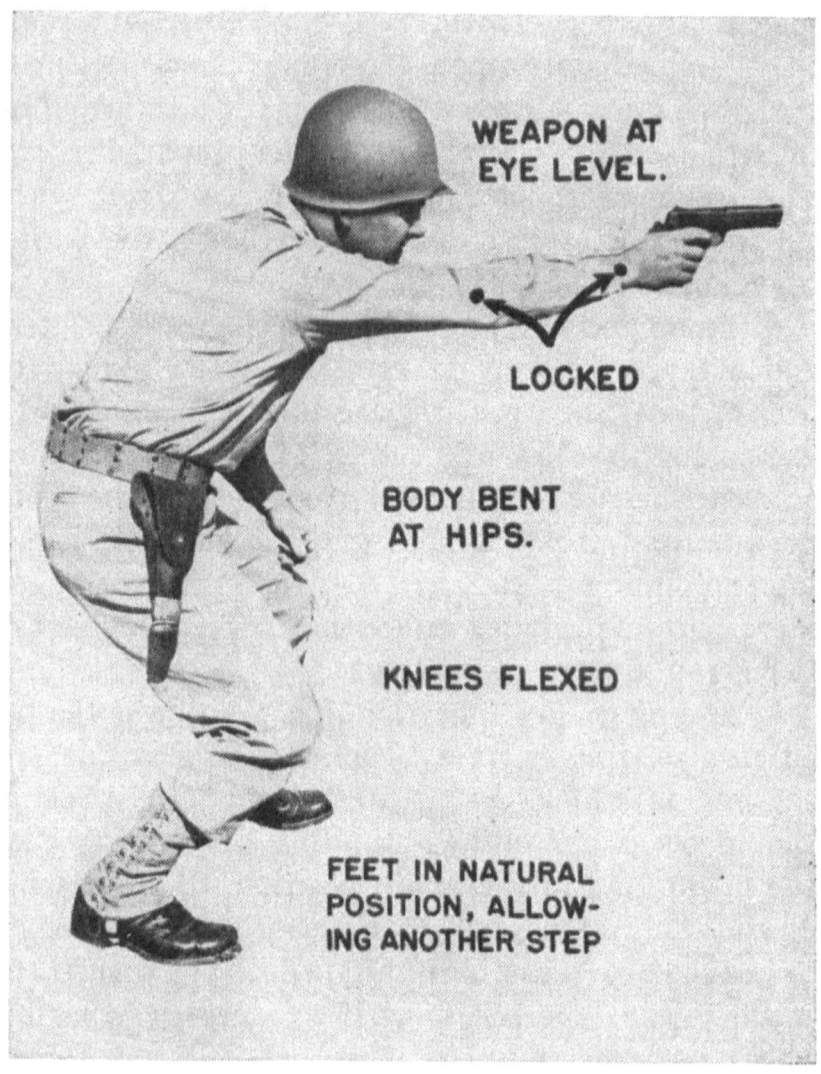

Figure 39. Crouch standing position.

(2) To practice the position, the firer thrusts the pistol at various targets at short ranges. *His eyes are directed at the target only.* Holding the pistol in position, he then drops his eyes and determines how closely his sights are aligned with the target. He then squeezes the trigger with a steadily increasing pressure and observes if his trigger squeeze moves the pistol out of alignment.

109. DISMOUNTED COMBAT FIRING RANGE. a. Range. Dismounted combat firing is conducted on the standard pistol range modified as described in following paragraphs. A suitable combat firing range may also be improvised from the standard rifle or submachine gun ranges and, with appropriate consideration for safety and other pertinent factors, other ranges may be improvised according to local circumstances.

b. Firing lines. Firing lines should be designated at about 50 yards, 25 yards, and 5 yards from the targets for prone, kneeling, and standing positions respectively. Safety factors will, in a measure, govern the distance of the firing line, for the standing position, from the line of targets.

c. Targets. (1) Standard silhouette targets (Target E) are used in dismounted combat firing and are controlled either from the pits or from behind the firing line. Since the target represents the figure of a man kneeling, interest is stimulated if a human figure is painted thereon. The use of pits may prove to be more satisfactory. If there is a shortage of personnel or equipment, stationery E targets on stakes could be used and the firing line controlled by voice or whistle. Considerable latitude must be given the range officers and officers in charge of firing.

(2) At least two targets, from 6 to 8 feet apart, should be prepared for each firing point. Each target should be independently controlled so that it may be hidden entirely from the firer's view or its edge turned toward him.

110. CONDUCT OF DISMOUNTED COMBAT FIRING. a. Preliminary. The object of combat firing is to familiarize the individual with positions not previously used and at the same time make the training more interesting and realistic. While it is intended that combat firing will be conducted without coaches, the officer in charge of firing should not hesitate to provide coaches for individuals who are in need of additional instruction. (See par. 93f.)

b. Combat firing course (fig. 40). (1) *Prone.* The firing order is formed at the starting line in rear of the 50-yard firing line, with firers opposite their respective targets. Upon command, weapons are loaded with 6 rounds, locked, and held at the ready position. The command 1. FORWARD, 2. MARCH, is given and the order moves forward at a walk. Upon arriving at the 50-yard firing line the command DOWN (fig. 37) is given and one target is exposed for 4 seconds. Two shots are fired at this target. At from 1- to 2-second intervals two more targets are successively exposed for 3 seconds each. One shot is fired at each target.

(2) *Kneeling.* Immediately upon completing the firing from the prone position and upon command, weapons are locked and the firing order advances.

Upon reaching the 25-yard firing line the command **KNEEL** is given. The kneeling position (fig. 38) is taken and weapons unlocked. From 2 to 3 seconds after the command **KNEEL** is given one target will be exposed followed by a second target at from 1- to 2-second intervals. The targets will be exposed for 3 seconds. One shot is fired at each target. Nine seconds later (14 seconds with the revolver) a third target will be exposed for 3 seconds. The weapons are reloaded with three rounds, and one shot is fired at this target within the 12-second interval (17 seconds with the revolver).

(3) *Standing.* Immediately upon completing the firing from the kneeling position and upon command, weapons are locked, and the firing order advances to about the 5-yard firing line. The crouch standing position (fig. 39) will be taken with weapons at the ready position, unlocked prepared to engage a target. One target will be exposed for 2 seconds followed at 1-second interval by a second target exposed for 2 seconds. One shot will be fired at each target.

RANGE:	50 YARDS	25 YARDS	15 YARDS	TWO 'E' TYPE TARGETS
POSITION:	PRONE	KNEELI	STANDING	6 TO 8 FEET APART
NO. OF ROUNDS TO BE FIRED:	4	3	2	TOTAL 9

Figure 40. Combat firing course.

(4) *Marking targets.* After completion of firing from the standing position, weapons are cleared, left on the 5-yard firing line and the firing order advances, marks scores, and pastes targets.

173

111. SCORING. The firer's individual score is equal to the total number of hits. The maximum possible score is 9 with either the pistol or revolver. Dismounted combat firing is not fired for qualification, but the following scores indicate the degree of proficiency achieved by individuals:

	Hits
Superior	7
Excellent	6
Satisfactory	5

112. SAFETY PRECAUTIONS ON THE RANGE. The safety precautions contained in paragraph 93 are applicable to dismounted combat firing.

Section III

MOUNTED

113. GRASPING WEAPON. The manner of grasping the revolver or pistol described in paragraph 79 is also applicable to mounted combat firing.

114. SIMULATED FIRING EXERCISES. With the targets set up as required for actual firing, the trooper simulates fire to the right front, right, right rear, left front, and left, bringing his horse to a halt upon the completion of each run. For firing mounted, it is of special importance that the pistol be pointed in prolongation of the forearm and not to the right of the forearm as is the common tendency. To enable men with average sized or smaller hands to avoid this tendency, the pistol barrel is swung well to the left even though only the tip of the index finger will then reach the trigger. These exercises should be made as realistic as possible. A perfunctory execution of them is a waste of time.

a. The trooper is given practice in simulating fire to the right (left) at the walk (figs. 41 and 42) and the gallop. From raise pistol the trooper points his pistol by thrusting it toward the target by the shortest possible line, not by flourishing the weapon over the arc of a circle. This movement should be steady, easy, and smooth. In bringing the pistol downward, he lowers the forearm in the direction of the target keeping the elbow slightly bent, the forearm and pistol in the same straight line, and the wrist rigid. He then leans his body well toward the target, resting the bridle hand on the horse's crest and, at the same time, forcing the right elbow well to the left, thus getting the elbow in the same vertical plane as the wrist and shoulder. He then extends the forearm so as to thrust toward the target, keeping the elbow forced well to the left. There are thus four movements: leaning the body toward the target, lowering forearm and pistol, forcing elbow to the left, and a thrusting toward the target. Throughout the exercise, he keeps his eyes fixed on the center of the target and concentrates on making hits (figs. 43 and 44).

b. The trigger squeeze is started at the beginning of the thrust and is continued so that, when the arm is fully extended or the thrust completed, the squeeze is completed and the piece is fired. Upon completion of the thrust, the target should be seen along the top of the pistol. If the next target is on the same side as the last the body remains leaning well toward the

Figure 41. Simulating firing to the right.

Figure 42. Simulating firing to the left.

Figure 43. Approaching target at raise pistol.

Figure 44. Simulating firing at target.

target and the elbow is bent ready for the next thrust. During the processes of firing, the bridle hand remains on the horse's crest.

c. In firing to the front (fig. 45), the trooper pushes his horse forward into the extended gallop and, with reins gathered up short, leans well forward and extends his pistol arm as far as possible to the front. To accomplish this he should half stand in the stir-

Figure 45. Simulating firing to the front.

rups. When the target is very low or close, the trooper extends his arm well to the front, being careful to have the muzzle of the pistol barrel well in front of the horse's face. He selects his target, fixes his eyes upon it with concentrated effort, catches sight of the target over his pistol barrel, and squeezes his trigger with steadiness and precision. He remains in this attitude simulating the firing of successive shots until he has passed the targets.

d. In firing to the rear (fig. 46), the trooper applies the principles of arm extension, seeing the target over the pistol, and squeezing the trigger as described in the preceding exercises. However, instead of half rising in the stirrups and leaning forward, he sits well down in the saddle and rotates his body to the

Figure 46. Simulating firing to the rear.

right and rear, being careful not to disturb his bridle hand nor to allow his lower legs to fly to the front. One of the principal causes for missing the target is failure of the firer to keep his eyes on the target until *after the shot is fired*. There is a tendency for the firer to become concerned with the next target with the result that he takes his eyes off the target he is engaging an instant before the shot is fired. The firer should prepare to engage the next target only *after* the shot is fired at the target being engaged.

115. TRAINING HORSES. a. Horses should be accustomed to the sight of barriers and targets and to the sound of firing. The training of horses is started by firing blanks in the vicinity of stables and corrals after horses have been worked or while they are feeding. The first shots, a few in number, should be fired some distance from the corral or stable. Each day this firing is brought closer and closer to the animals until shots are finally fired in the air directly in the midst of the horses. Horses stand for this firing better in groups than they do singly.

b. After the horses have become thoroughly accustomed to the firing under these conditions, they are taken out with a rider and an individual on the ground, both armed with a pistol. The rider points the pistol in various directions and snaps the trigger while the man on the ground fires blanks. At first the dismounted man places himself a short distance from the horse, gradually moving closer as he fires until the horse no longer pays any attention to the movement of the gun, the click of the trigger, or the firing of the shot.

c. Firing is then started from the animal. For the first few periods, very few shots are fired and these always to the rear. Firing these initial shots to the rear is essential to good training since the average horse objects to the firing only because of the noise and muzzle blast in his ears. Therefore, he must be accustomed to the noise by gradually changing the direction of fire from rear to front.

d. In firing to the front care must be taken to place the pistol well in advance of the horse's ears (fig. 45), thereby preventing the noise from going directly into the ears and annoying the horse.

e. Horses should also be accustomed gradually to the strangeness of the targets used in the prescribed courses. A horse that shows fear of targets should not be forced roughly up to them, but should be patiently ridden in the vicinity of the targets, always approaching them slowly and easily, until he becomes accustomed to passing close by any target without fear.

f. If the above methods are used and the trainer is patient and willing to go slowly, most horses respond readily to this training.

116. CONSTRUCTION OF COURSE. The course is laid out as shown in figure 47. The length of the course measured from start to finish is 345 yards. The area should be level or with only a gentle slope. Points on the course are marked with flags or posts as indicated on the diagram. The curves, A, B, and C, have a 10-yard radius. The other curves have a radius of about 20 yards. Targets are of the E and the M type and are located as shown in figure 47 by the numbers 1 to 7. Barriers at least 18 inches in height are placed as indicated by the heavy lines. A flag is placed at the starting point and at the finish line as indicated.

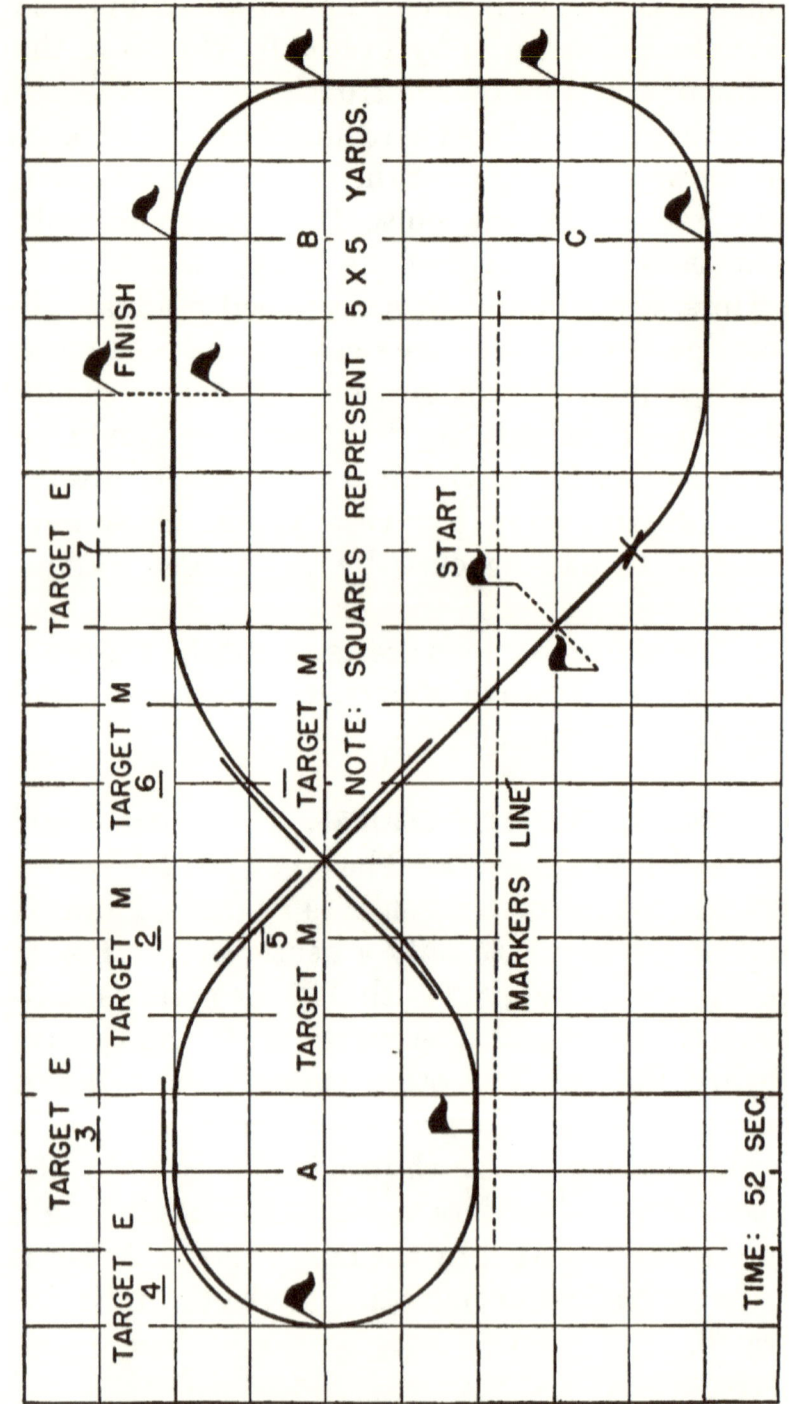

Figure 47. Mounted pistol range.

117. MOUNTED COMBAT FIRING COURSE. The following table indicates the minimum firing to be done on the mounted combat firing course.

	Runs twice around the course	Targets	Range	Gait	Time limit
Firing to the right front, right, right rear, left front, and left.	2 runs of 14 shots each.	4 M 3 E	5 to 7 yards.	Gallop..	52 seconds for each run (twice around the course).

118. MODIFICATION FOR FIRING REVOLVER. The course prescribed for the automatic pistol should be modified as follows to accommodate the revolver:

a. Fire at the first six targets only.

b. Scores are four points less than that prescribed for the automatic pistol.

c. Additional time is given (if necessary) for the purpose of reloading.

119. CONDUCT OF MOUNTED COMBAT FIRING. a. General. (1) *Marking and scoring.* The targets are designated as shown in figure 47. The number of men detailed as markers is left to the discretion of the officer conducting the firing. The markers take position approximately opposite their targets and on the designated line. When a complete score has been fired, the markers run to their targets, examine them, face the scorer, and call in their numerical order, the hits or misses. For example: "No. 1, a hit," "No. 2, a miss," or "No. 3, two hits." After calling the hits or misses, the markers cover shot holes with pasters and run back to their positions.

One noncommissioned officer, detailed as scorer, is posted behind the markers at a convenient place to hear the calls.

(2) *With the revolver.* Procedures described herein apply to firing the automatic pistol and, where practicable, they also apply to firing the revolver.

b. Value of hits. Each hit on the target counts one point.

c. Time limit. The time limit for each run is 52 seconds. The time is measured from the instant the firer passes the flag at the starting point to the instant he passes the flag at the finish point. In case of a defective cartridge or malfunction the firer is allowed sufficient time for the correction thereof.

d. Horses. So far as practicable, individual firers are required to ride the horses regularly assigned to them. In no case is one horse to be used by more than two individuals nor are more than 25 percent of the individuals of a troop permitted to fire from horses other than those regularly assigned to them.

e. Firing. (1) The firer is equipped with pistol, lanyard, field belt, and two loaded magazines of seven rounds each. Just prior to starting a run, he inserts a loaded magazine, loads and locks the pistol, and takes the position of raise pistol. The other magazine remains in the magazine pocket. Starting at the gallop at X, the firer rides past the starting point, unlocks his pistol, and rides around the course as indicated in figure 47, firing one shot at each of the targets, Nos. 1 to 7, as he encounters them. When he has fired his seventh shot, the firer, continuing around the course at the gallop, changes magazines, loads and locks the pistol, and takes the position of raise pistol. As he passes the flag at the starting point the second time, he unlocks his pistol and fires again at targets Nos. 1 to 7 as each is encountered. Upon changing magazines after passing target No. 7 the first time, the firer may place the empty magazine in any secure place most convenient to him, such as a shirt pocket, inside the shirt or, if he so desires, he may return the magazine to the magazine pocket.

(2) Prior to making a complete run as outlined in (1) above, the trooper should be afforded practice in going through the course and simulating fire on all targets, and going through the course firing blank cartridges from the caliber .45 pistol at six targets selected at random.

(3) For additional practice the firer may make runs firing ball cartridge from one magazine and simulating fire from an empty magazine for the other part of the course. Other variations, such as loading less than seven rounds in a magazine and firing at certain designated targets, may be prescribed by the organization commander.

(4) Penalties are exacted as follows:

(*a*) For each 5 seconds or fraction thereof in excess of the prescribed time limit of 52 seconds, one point is deducted from the final score.

(*b*) For each instance during a run that the horse assumes a gait other than the gallop, one point is deducted from the firer's score. This penalty is not to be construed as applicable to a case where the horse in changing leads hits two or three beats of the trot.

120. SCORING. The firer's individual score is equal to the total number of hits less deductions explained in previous paragraph. The maximum possible scores are 28 for the pistol and 24 for the revolver. Mounted combat firing is not fired for qualification, but the following scores indicate the degree of proficiency achieved by individuals:

| | Scores ||
	Pistol	Revolver
Superior	22	19
Excellent	20	17
Satisfactory	17	14

121. SAFETY PRECAUTIONS ON THE RANGE. a. Safety precautions. In order to teach the trooper to handle his pistol with safety, he is frequently required to go through the motions of loading and unloading, locking and unlocking the pistol while mounted, at the halt, and at all gaits. The trooper should also be practiced in withdrawing and inserting magazines at all gaits. During this preparatory instruction, the trooper must be taught by exercises in simulated fire to observe the following rules (this should be done daily until the proper movements in their proper sequence become matters of habit and are instinctively performed when they become necessary):

(1) Always lock the pistol after loading and keep it locked until about to fire.

(2) When a cocked pistol is held in the hand, it

must always be held at raise pistol until it is locked, or until it is necessary to fire, load or unload. If it becomes necessary to lower the pistol for any other purpose, the pistol must first be locked. Mounted men should never under any circumstances use both hands on the reins when the pistol is drawn.

(3) At the slightest misbehavior of the horse, the pistol must be locked. The trooper then makes a fresh start to complete his firing. If the horse rears, plunges, bolts, stumbles, or leaves the track, the trooper should instinctively lock the pistol.

Section IV

COMBAT FIRING TARGETS

122. TARGET E. Target E is a drab silhouette, representing a kneeling figure, amde of bookbinders' board or similar material (fig. 33).

123. TARGET M. Target M consists of two parts: The upper part is Target E and the lower part is a trapezoidal piece whose upper edge is placed closely against the lower edge of Target E. It is made of material similar to Target E (fig. 48).

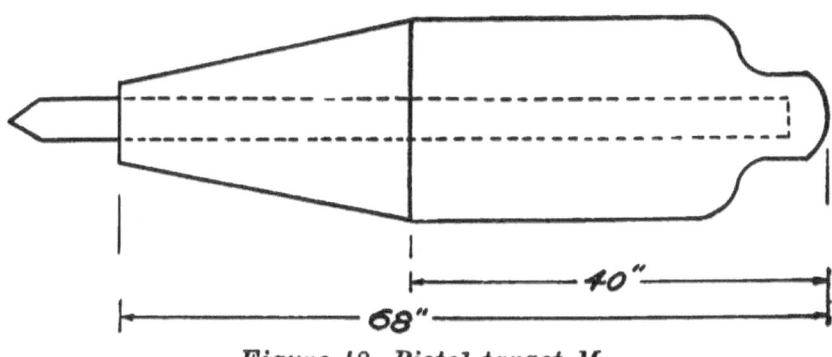

Figure 48. Pistol target M.

Section V

SMALL-BORE PRACTICE

124. GENERAL. The general matter contained in paragraphs 100 to 104, inclusive, applies equally to dismounted and mounted combat firing. Mounted pistol practice with the caliber .22 pistol is of great value in accustoming the horses to the sound of firing from the saddle and to running the mounted course. The lesser noise and concussion does not distract them as does the caliber .45 pistol. The lesser cost of ammunition permits more instruction practice than with caliber .45.

125. SMALL-BORE RANGES. The ranges used for small-bore dismounted and mounted combat firing are the same as prescribed in paragraphs 109 and 116.

126. CONDUCT OF SMALL-BORE COMBAT FIRING. Small-bore dismounted and mounted combat firing is conducted as prescribed in paragraphs 110 and 119.

CHAPTER 3

ADVICE TO INSTRUCTORS

Section I

GENERAL

127. PROVISIONS NOT MANDATORY. Unless specifically designated as such, information and suggestions contained in this chapter are not mandatory. They are intended as a guide for use by personnel responsible for the instruction of troops in subjects covered herein.

128. METHOD OF INSTRUCTION. It is advisable to make use of the applicatory system where instruction is conducted in subjects covered in this manual. Comprising this system of instruction are several phases; explanation, demonstration, application (practical work), and examination.

   a. Explanation. The initial explanation and demonstration of any particular phase of the instruction is presented to the assembled unit by the instructor assisted by essential demonstration personnel. The general purpose of the entire course or period of instruction should be explained first. The various phases or steps of the course should then be presented in a series of explanations and demonstrations.

   b. Demonstration. (1) Demonstrations which are skillfully conceived and executed expedite and simplify instruction as well as stimulate interest. Successful demonstrations are usually short and concise. They leave the student with an exact impression stripped of superfluous details. The demonstrations incident to all subjects should be arranged in progressive sequence, and where practicable should alternate with practical work to permit the student to fix these successive phases of instruction in his mind.

(2) The men who constitute the demonstration unit should be carefully selected for their intelligence, ability, and appearance. They should be thoroughly trained and rehearsed in the duties they are to perform so that the demonstration will proceed smoothly and illustrate clearly and simply the phase of instruction being presented.

(3) The equipment used for demonstrations should be the best available. A demonstration platform or an area in which the students can be assembled quickly at a position from which they can see and hear every part of the demonstration is essential.

(4) Interest is added and valuable instruction given by repeating demonstrations, including common errors, and requiring the students to detect these errors.

c. Application (practical work). (1) This third step of instruction is of major importance since it gives the student an opportunity actually to accomplish that which has been previously explained and demonstrated.

(2) During the practical work phase of instruction, best results are obtained if the unit is divided into groups. Groups should consist of from four to eight men depending upon the number of men undergoing instruction and the number of assistant

instructors available. Each group is provided with a set of equipment and placed under the direct supervision of a trained assistant instructor. The group then executes the previously demonstrated phase of instruction, individuals rotating within the groups, until all men have mastered the instruction.

(3) The initial allotment of time and equipment should be made carefully. However, the instructor should not hestitate to alter this allotment if the majority of the men fail to master the instruction within the allotted time or are kept at one exercise to the point of boredom. The frequent rotation of duties within each group is preferable to keeping each man in one position for a long time.

d. Examination. An informal oral or practical examination should be conducted upon completion of each phase of instruction. In addition to the examination required before starting range practice, the organization commander should conduct such additional examinations as are necessary to insure that all men have completed the training.

Section II

MECHANICAL TRAINING

129. GENERAL. The unit to be instructed is assembled in a suitable area and divided into conveniently sized groups, each under the supervision of an assistant instructor. The instruction is centralized under the supervision of the unit instructor. Explanation and demonstration are concurrent, each assistant instructor demonstrating the elements of the particular phase of instruction as the instructor explains it from the platform. For short periods of practical work the instruction is decentralized under the assistant instructors.

130. DISASSEMBLY AND ASSEMBLY OF THE PISTOL.
a. Equipment required. One pistol (revolver) with magazine (2 clips) per man, one pistol cleaning kit per group.
  b. Procedure. An assistant instructor disassembles and assembles the pistol (revolver) while the instructor is explaining the procedure.
  (1) *Practical work.* Assistant instructors explain and demonstrate the procedure and each student performs each operation in unison with the assistant instructor. When acquainted with the procedure, each student disassembles and assembles the pistol (revolver) without assistance.
  (2) *Questions and answers.* Students ask questions concerning points about which they are doubtful; instructors ask questions designed to test the effectiveness of their instruction.

131. CARE AND CLEANING. a. Equipment required. Same as described in paragraph 130 above, plus additional equipment required for demonstration purposes.
  b. Procedure. (1) The pistol (revolver) is compared with any other mechanism and the necessity explained of keeping it clean, lubricated, and in effective condition.
  (2) Concurrently, the proper method of cleaning the pistol is explained and demonstrated.

(3) Instruction in cleaning is conducted under supervision of squad and platoon commanders throughout the year.

(4) Students ask questions concerning points about which they are in doubt; instructor asks questions designed to test effectiveness of his instruction.

132. FUNCTIONING. a. Equipment required. Same as described in paragraph 130.

b. Procedure. (1) The various phases of loading, unloading, and firing the pistol (revolver) are explained.

(2) Students practice loading a magazine (clips), inserting the magazine (clips) in the pistol, and loading and unloading the pistol (revolver).

(3) Students ask questions concerning points about which they are in doubt; instructor asks questions designed to test effectiveness of his instruction.

133. ACCESSORIES. a. Equipment required. One each of the accessories listed in paragraph 16 or 61.

b. Procedure. (1) The use of each accessory is explained and demonstrated.

(2) Students examine accessories.

(3) Questions are asked by both student and instructor as described in paragraph 132.

134. INDIVIDUAL SAFETY PRECAUTIONS. a. Equipment required. Same required in paragraph 130.

b. Procedure. (1) Safety rules discussed in paragraph 24 or 63 are explained and demonstrated. Tests for safety devices are explained and demonstrated.

(2) Students practice the application of the safety

rules discussed and the tests of safety devices. Observation of the rules for safety becomes a habit only after constant practice over a long period. Instructors and assistant instructors must be constantly alert to enforce the safety rules.

(3) Questions are asked as described in paragraph 132.

Section III

MANUAL OF THE PISTOL AND REVOLVER

135. GENERAL. a. Instruction in the manual of the pistol and the revolver is conducted concurrently with dismounted and mounted drill and with previous instruction in this chapter.

b. Manual of the pistol and manual of the revolver lend themselves readily to the applicatory system of instruction.

c. Equipment required. Each man is equipped with a pistol with magazine (revolver), holster, belt, magazine pocket, two extra magazines (clips), and pistol lanyard (for mounted instruction).

d. Procedure. (1) The instructor, employing a trained demonstration unit, explains and demonstrates each movement in the manual.

(2) Under assistant instructors, groups are drilled in the various movements.

(3) Each group is tested by the instructor at the end of each period and a critique is conducted.

Section IV

MARKSMANSHIP

136. GENERAL. a. Marksmanship is the basic step in training the soldier to employ successfully the pistol or revolver in combat. A soldier will subconsciously employ in combat the principles he has been taught in marksmanship, hence these principles must be sound.

b. The procedure used in conducting marksmanship instruction is similar to that used in the preceding sections of this chapter except that it is more decentralized. During instruction in preparatory exercises, the entire unit is assembled initially under the unit instructor who is assisted by a trained demonstration unit. Following the initial explanation and demonstration, the groups move to their individual sets of equipment and start practical work under the assistant instructors.

c. Firing exercises should be conducted under centralized control.

137. PREPARATORY RANGE TRAINING. a. General. (1) A thorough course in preparatory range training is essential. During this period, the soldier learns all the mechanics of target practice except actual firing. Preparatory training may be done in barracks. Range equipment is not required.

(2) Adequate time should be allowed and thorough supervision provided to insure that each man has thoroughly mastered the instruction before he is permitted to fire.

(3) Each instructional phase is taken up in proper sequence and training in that step is completed by each man before the next step is initiated. If men fail to progress uniformly, groups should be rearranged so that instruction will not be held up by men who are slow to learn.

(4) A careful record should be kept of the progress of each man and each group in order that the instructor will know the progress of instruction and when the men are ready for range practice.

b. Equipment per group. (1) One sighting bar.

(2) One pistol (revolver) rest.

(3) Two small aiming disks.

(4) One 5-inch aiming disk.

(5) Two small boxes with paper tacked on one side.

(6) One target frame on which is placed a blank sheet of paper at least 2 feet square.

(7) One target L.

(8) One target E, bobbing.

(9) One pistol with magazine (revolver) holster, belt, magazine pocket, and two extra magazines (clips) per man.

(10) Material for blackening sights.

(11) Tissue paper for copying shot groups.

(12) Pencils.

(13) Additional equipment such as blackboard, charts, and drawings as decided by the instructor.

c. Procedure. (1) Each phase of training is initiated with an orientation discussion covering the purpose and scope of the instruction concerned and the manner in which it is to be conducted. Where appropriate, the instructor employs a trained demonstration group to emphasize important points of instruction.

(2) When practical work is undertaken, groups are kept as small as circumstances permit and are supervised by assistant instructors.

(3) The proficiency of the students is determined by means of oral examination and by close observation of the manner in which they accomplish practical work.

138. RANGE FIRING. a. General. The number of men who are to fire and the availability of range facilities are the primary factors affecting details of administration and supply. These matters must be the subject of detailed prior planning in order that firing may proceed smoothly and without interruption. Plans should include training of necessary assistants and demonstration units and adequate provision for keeping all men occupied by concurrent marksmanship training of an appropriate nature.

b. Equipment. (1) *Pistols (revolvers)*. Every effort should be made to detect and repair mechanical defects prior to conducting marksmanship training.

(2) *Magazines*. Damaged or inoperative magazines are the greatest single cause of malfunctioning of the pistol. Principal defects in magazines are dents, spread lips, and the presence of sand or dirt below the follower.

c. Safety precautions. Range officers, the officer in charge of firing, and the commander responsible for the location of ranges and the conduct of firing thereon must have a thorough knowledge of AR

750–10. Before firing is commenced, all officers and men who are to fire or who are concerned with range practice will be familiarized with the safety precautions contained in paragraph 93.

139. CONSTRUCTION OF TARGETS AND RANGES. a. General. For detailed information relative to targets and target accessories, see T/A 23, Targets and Target Equipment.

b. Targets. When regular printed targets are not available, suitable substitutes can be made on sheets of wrapping paper. Dimensions should be accurate. Targets can be made in large numbers by using a stencil made of heavy linoleum.

c. Ranges. (1) *General.* The range should be level and open and if practicable, so located that fire can be delivered against a steep hill or bank in rear of the targets. Semipermanent bases should be constructed to facilitate placing targets in position and changing targets with the minimum delay or confusion.

(2) *Marksmanship course.* Targets should be spaced from 3 to 5 yards apart. The depth of the range should be not less than 40 yards.

(3) *Combat firing courses.* Dimensions outlined in paragraphs 109 and 116 should be accurately observed.

INDEX

	Paragraph	Page
Accessories:		
Automatic pistol	16, 133	25, 193
Revolver	61, 133	74, 193
Ammunition	17–23	26, 29
Application, method of instruction	76c, 128c	87, 190
Arms, manual of:		
Automatic pistol	26, 36, 135	33, 39, 194
Revolver	65, 71, 135	77, 81, 194
Assembly of—		
Automatic pistol	4, 130	10, 192
Revolver, Colt	46, 130	52, 192
Revolver, Smith and Wesson	48, 130	55, 192
Ballistics, automatic pistol	2	3
Bar, sighting	78a	88
Calling the shot	92e	142
Cards, score	96q	155
Care, ammunition	22	29
Care and cleaning:		
After chemical attack	10, 54	18, 63
Automatic pistol	5–7, 131	15, 192
On the range	8, 52	17, 63
Points to be observed	11, 55	19, 63
Revolver	49, 50, 131	60, 192
Under unusual climatic conditions	9, 53	17, 63
Cartridge:		
Dummy, use	92b	138
Markings	21a	27
Types and models	21b	27
Chamber, automatic pistol:		
Close	30	36
Mounted	36	39
Open	29	34
Mounted	38	39
Classification, ammunition	18	26
Coach and pupil	74	82

	Paragraph	Page
Coaching:		
Quick fire	92g	143
Range practice	92a	137
Slow fire	92c	138
Sustained fire	92g	143
Timed fire	92f	142
Trigger squeeze	92d	140
Cocking, revolver:		
Side-method	83c	113
Straight-back method	83c	113
Combat firing:		
Course	110b	172
Definition	105a	164
Range	109	171
Computing scores	94, 111, 120	144, 174, 186
Conduct, range practice:		
Combat firing	110	172
Dismounted	92	137
Mounted	119	183
Construction:		
Mounted firing course	116	181
Principles governing	99	160
Ranges	98, 139	159, 198
Targets	139	198
Coordination of movement	108	169
Course:		
Combat firing	110b, 117	172, 182
Mounted firing, construction	116	181
Practice, small-bore	103	163
Small-bore, mounted	116	181
Danger signals, range	99g	161
Data:		
Automatic pistol	2	3
Revolver	44	45
Demonstration	76b, 128b	87, 189
Description:		
Automatic pistol, M1911 and M1911A1	1	1
Revolver, Colt and Smith & Wesson	43	41

	Paragraph	Page
Devices, safety:		
Automatic pistol	13	21
Revolver	57	65
Tests	25, 64	32, 76
Disassembly:		
Automatic pistol	3, 130	6, 192
Revolver, Colt	45	46
Revolver, Smith & Wesson	47	53
Disks, aiming	78	88
Equipment, aiming exercise	78a	88
Examination:		
Automatic pistol	85, 128d	122, 191
Revolver	86, 128	127, 189
Exercises:		
Aiming	78	88
Cocking, revolver	83d	117
Position	79	96
Timed fire	81	106
Sustained fire:		
Automatic pistol	82	108
Revolver	83	111
Simulated firing, mounted	114	174
Trigger-squeeze	80	102
Explanation, method of instruction	128a	189
Factors, safety (on range)	98c	159
Fire:		
Quick, training for	84	120
Timed	81	106
Sustained:		
Automatic pistol	82	108
Revolver	83	111
Firing:		
Combat:		
Course	110b	172
Definition	105	164
Mounted	119	183
Mounted, modification for revolver	118	183
Range	138	197

	Paragraph	Page
Functioning:		
Automatic pistol	14, 132	22, 193
Revolver, Colt	58, 132	66, 193
Revolver, Smith & Wesson	59, 132	69, 193
Grade, ammunition	20	27
Grip:		
One-hand	106a	165
Two-hand	106b	165
Horses:		
Training	115	180
Use in combat firing	119d	184
Identification, ammunition	21	27
Instruction:		
Conduct	76	86
Individual, importance	74c	83
Methods	74	82
Instructors	74e	84
Magazine, automatic pistol:		
To insert	31	37
To insert (mounted)	39	39
To withdraw	28	34
To withdraw (mounted)	37	39
Marksmanship, training	136	195
Objective	73	82
Scope	73	82
Methods:		
Coaching:		
Range practice	92a	137
Slow fire	92c	138
Instruction	74, 128	82, 189
Movement, coordination of, combat firing	108	169
Number, lot, ammunition	19	26
Objective:		
Combat firing	105b	164
Practice, small bore	100	162
Preparatory training	75a	86
Training, marksmanship	73	82

	Paragraph	Page
Operation:		
Automatic pistol	12	20
Revolver	56	65
Organization, on range	94	144
Parts, spare:		
Automatic pistol	15	25
Maintenance, revolver	60b	74
Organization, revolver	60a	74
Pistol:		
Manual of	26, 36, 135	33, 39, 194
To fire	1	1
To grasp	79a, 106, 113	96, 165, 174
To inspect	34	38
To inspect (mounted)	42	40
To load	32	38
To load (mounted)	40	40
To raise	27	34
To return	35	38
To unload	33	38
To unload (mounted)	41	40
Position:		
Combat firing:		
Kneeling	107b	167
Prone	107a	166
Ready	107d	168
Standing	107c	168
Exercise	79	96
Practice:		
Additional, small bore	104	163
Combat firing, mounted	119	183
Instruction	88	132
Marksmanship	74d	84
Range, combat firing, conduct	109	171
Range	91	136
Record	89	134
Small bore	100, 124	162, 188
Precautions:		
Individual, safety	134	193
Safety, on range	93	143
Principles, sustained fire, revolver	83	111

	Paragraph	Page
Quick fire:		
Coaching	92g	143
Course prescribed	89c	135
Training	84a	120
Range:		
Organization	94	144
Preparatory training	137	195
Ranges:		
Combat firing	109, 116, 125	171, 181, 188
Construction	99, 116, 137c	160, 181, 196
Rules for selection	98b	159
Safety factors	98c	159
Size	98f	160
Rate of fire, revolver	43	41
Record practice	89	134
Regulations governing	96	152
Rest, automatic pistol and revolver	78a	88
Revolver:		
Cocking	83c	113
Manual of	65	77
Modification of firing course, mounted	118	183
To fire, double action	43	41
To fire, single action	43	41
To inspect	69	80
To inspect, mounted	72	81
To load	67	78
To raise	66	78
To return	70	80
To unload	68	80
Types	44	45
Safety:		
Individual, precautions	134	193
On range	93, 112, 121	143, 174, 186
Rules for	24	30
Revolver	63	75
Scope:		
Combat firing	105c	164
Preparatory training	75b	86
Training, marksmanship	73b	82

	Paragraph	Page
Scoring	90, 111, 120	135, 174, 186
Sights, blackening	77	87
Signals, danger, on range	99g	161
Slow fire:		
Course prescribed	88	132
Coaching	92c	138
Firing	94e	145
Storage, ammunition	23	29
Supervision, individual instruction	74c	83
Sustained fire:		
Automatic pistol	82	108
Coaching	92g	143
Course prescribed for	88c, 89b	133, 134
Firing	94e	145
Revolver	83	111
Target details	95	151
Targets	97	156
Combat firing	109c, 122, 123	171, 187
Construction	139b	198
Small bore	97d	159
Tests, safety devices	25, 64	32, 76
Timed fire:		
Automatic pistol	81	106
Coaching	92f	142
Course prescribed	88b, 89a	133, 134
Firing	94e	145
Training:		
Horses	115	180
Marksmanship, objective	73	82
Mechanical	129	191
Preparatory	75	86
Range	137	195
Quick fire	84	120
Sustained fire	82, 83	108, 111
Timed fire	81	106
Trigger squeeze, importance	80	102

www.ingramcontent.com/pod-product-compliance
Lightning Source LLC
Chambersburg PA
CBHW031107080526
44587CB00011B/865